ABOUT THE AUTHOR

Carol Ann Knitzer has taught elementary, intermediate, high school and junior college in New York and Florida. Holder of a B.S. in Mathematics and an M.A. in Math and Education from Brooklyn College, City University of New York, she has presented workshops for parents and teachers interested in using games to teach mathematics. She has also conducted math labs for groups of remedial students.

4

ISBN 0-13-547745-X

MAKING MATHEMATICS FUN!

Learning Games and Enrichment Activities
for the Elementary Classroom

MAKING
MATHEMATICS
FUN!

Learning Games and Enrichment Activities
for the Elementary Classroom

Carol Ann Knitzer

Parker Publishing Company
West Nyack, New York

Library of Congress Cataloging in Publication Data

Knitzer, Carol Ann.
    Making mathematics fun.

    Includes index.
    1. Mathematical recreations. 2. Mathematics—
Study and teaching (Elementary) I. Title.
QA95.K54   1983   372.7   82-22290
ISBN 0-13-547745-X

Printed in the United States of America

About the Author

Carol Ann Knitzer has a diverse background in mathematics education, having taught elementary, intermediate, high school and junior college in New York and Florida. She has worked as a math specialist and has presented workshops for parents and teachers who are interested in using games to teach mathematics. She has also conducted elementary math labs for groups of remedial students and has developed many of her own materials for teaching math fundamentals.

Ms. Knitzer holds a B.S. in Mathematics and an M.A. in Math and Education from Brooklyn College, City University of New York.

To Philip, Laura, and Daniel

Acknowledgements

The author wishes to express her gratitude to all those in the education community who shared their ideas, and gave their assistance and encouragement toward the development of this book.

A Word on This Book's Practical Value

This book is for teachers who believe that the learning of mathematics should be fun, and that it can be stimulating, challenging, and enjoyable—not just for some students, but for all students. The wide range of mathematical games and activities in this book represents a simple and effective means toward this end, whether your objectives involve the mastery of essential arithmetic facts, or the understanding of key mathematical concepts.

Many of these games are completely original, developed by the author, and enable you to teach or review a specific skill or introduce a new idea. Some are variations of pre-existing games with an interesting mathematical twist. All games and activities have one common concern—how to develop basic skills and concepts, while at the same time building positive attitudes toward mathematics.

There is no single system that will teach mathematics successfully to all children. To provide students of varying abilities with an effective mathematics program, we must offer a diversity of materials and approaches. This book, when used as a complement to the current math curriculum, will allow teachers to present a wide range of different approaches, enabling each student to work at a level consistent with his or her own interests and abilities.

Each area of mastery is presented in a variety of thoughtfully developed ways, ranging from enjoyable concrete activities that allow children to make intuitive discoveries, to exciting and stimulating games involving abstract concepts. For example, children can learn the necessary skills and concepts of subtraction through involvement with dice, cards, gameboards, spinners, and puzzles. And while working with ancient learning devices, students will have the opportunity to discover and understand the concept of regrouping. Fractions can be introduced concretely through an activity that allows pupils to manipulate fractional parts, followed by a series of games designed to develop proficiency in working with fractions and decimals. Balancing out the picture is a look at very early fractions in a section on mathematics throughout history.

Meaningful variations are suggested: a card game for reviewing addition facts, sure to be a favorite of first and second graders, can be converted for upper elementary students to introduce the need for

9

common denominators; a sport activity involving subtraction might be changed over to multiplication and division. The materials presented here go beyond the basics to include historical approaches, puzzles, tricks, geometry, logic, and creativity.

All of the math games, activities, and variations in this book are cross-referenced and listed by specific skills and concepts to be taught. Overviews are provided to allow for the selection of materials that satisfy the immediate needs of the class. Many of these games and devices will prove to be so popular that additional sets should be made and a procedure established permitting the games to be taken home and shared with parents and friends.

These diverse materials can be used effectively in all types of classroom situations and are adaptable to any teaching style. They have been successful as whole class activities, for large and small groups, in learning centers, and in resource rooms. Using these materials provides a setting that allows for student interaction and cooperation, enhances students' self-esteem, and creates a more satisfying classroom atmosphere.

Carol Ann Knitzer

How to Use This Book

Initially, you might wish to browse through these chapters and familiarize yourself with the full range of activities that are available. The materials in this book are meant to be used selectively—allowing you to choose the games and activities most appropriate for your students. To efficiently select the most suitable materials, there are three principal sources you should be familiar with: Table of Contents, Subject Index, and individual Overviews.

The *Table of Contents* presents the basic organization of the book and lists each activity, game, or topic included within the chapters. While the first four chapters are organized around specific areas of mathematics (numeration, addition, multiplication, and fractions), these areas are also interspersed in the subsequent chapters.

Therefore, the *Subject Index* at the end of this book would be your most valuable source for finding all games and activities involving a specific area. Perhaps this is best understood by means of an illustration. Let us take the area of multiplication. Chapter 3 contains permanent multiplication games. Each of these could be made in advance, then used numerous times by the entire class. However, this chapter is not the only section containing multiplication activities. The *Subject Index* would also refer you to quick and easy games in Chapter 5, historical activities in Chapter 7, and tricks involving multiplication in Chapter 8. In addition to all this, Chapter 6 contains 20 varied and adaptable games and activities requiring little or no preparation. Finally, also under multiplication, you would be referred to Chapter 10 for appropriate card games.

The *Subject Index* would also list games presented originally in a different subject area, for which a variation using multiplication has been suggested. Of course numeration, addition, fractions, and other topics are also treated in the same manner.

Which of these activities will meet the immediate needs of your students? Do you need a game for the whole class, or an activity for a small group? Should this be a game or a project that might be prepared in advance, or do you prefer a series of quick and easy games that require no preparation?

To enable you to understand exactly what each game and activity has to offer, and how they might meet the specific needs of your classroom, *Overviews* have been provided in the appropriate chapters. *Overviews* are first offered at the beginning of the chapter and also immediately preceding each activity.

PREPARING MATH GAMES

As you become familiar with the organization of this book, the diversity of materials and approaches becomes apparent. The book includes:

- permanent games for small groups
- quick and easy games for the whole class
- historic learning devices and methods
- investigative activities and challenging problems
- projects for individual students

The materials needed for these games and activities are readily available. They include scissors, pens, tagboard, tape, and index cards. A good portion of these activities need only paper and pencil or a chalkboard. The required supplies are listed in advance of each game, activity, or project.

If permanent games are desired, gameboards can be easily made out of tagboard, oaktag, posterboard, or file folders. Making games colorful and attractive will add to their appeal. Both posterboard and index cards are available in a variety of colors. Finally, games can be covered by clear contact paper for durability and protection.

To add the element of chance to some of the games, dice or spinners are suggested. Standard dice can be used, or soft (and quiet) dice can be made in the following manner: Cut a sponge or a piece of styrofoam into cubes and apply numerals, either by writing or pasting them on—depending on the materials used. Care must be taken so that the numbers will not come off when dice are held. Spinners are easily made by using paper plates, brass fasteners, and paper clips. Specific step-by-step instructions are given with each game. Since making the games can be just as exciting as playing them, allow your students to become involved in their preparation whenever possible.

Be flexible. If you or your students have any ideas on how to improve a game, go right ahead. Make any changes necessary to allow the game to better meet the needs and objectives of your classroom.

GAMES FROM DIFFERENT CULTURES

A number of the math games offered here are based on games from different cultures. Old, standard games that were popular in other countries and in the past were studied. By modifying the rules, and by introducing mathematics, a new math game was created. While the new game still maintains the essence of the original, it now becomes more educationally useful. For example:

Indian Stones (Chapter 2) converts a game that was originally played by native Americans into a team game where addition combinations are reviewed.

Spanish Subtraction (Chapter 2) was based on an ancient Spanish game; but now the game has been modified and at every turn the directions are controlled by subtraction.

Roman Chase (Chapter 3) was suggested by a game originally from ancient Italy. In this version, the players who know their division facts are able to obtain extra turns.

Fraction Bean Pot (Chapter 4) uses an old Hebrew game and converts it into a group game where players take fractional parts of a set of objects. Although beans are suggested, using edibles such as candy and nuts can add to the fun.

The Jungle Game (Chapter 4) is perhaps the most complex and difficult game in the book. It was based on a Chinese game similar to chess, but in our variation it is played like this: each time a piece is captured, a fraction must be converted into a decimal. Since many students have not yet advanced into the study of decimals, a variation is suggested in which only the knowledge of multiplication is needed.

African Stone Game (Chapter 9) is a game originally from Africa and which presently has many variations. While not containing numbers, the game involves counting, quantitative thinking, and logic. Rules were selected from different variations which combine to form a game suitable for most students.

COMPETITION AND MOTIVATION

Competition: is it necessary? Many of the games here are presented in a competitive manner with directions that refer to teams and game winners. Competition can be a force for good, keeping the game going and interest strong.

However, sometimes competition might go against the tone of the classroom as established by the individual teacher; some teachers prefer to have students compete only with themselves. Under these circumstances, the competitive aspect of the games can be minimized or eliminated.

   In either case, immediately preceding a game is an excellent time for reviewing the specific skills that will be needed to play. Be sure students understand the relationship between these skills and the game to be played. If the review is short and to the point, it will not take away from the students' enjoyment of the game, but will add to the game's relevance.

Table of Contents

MAKING MATHEMATICS FUN!

Learning Games and Enrichment Activities
for the Elementary Classroom

CHAPTER **1**

The Beginning—
Counting, Numerals and Place Value

Throughout the preschool years, our students have been learning through play. They have taken their playful activities and games very seriously, worked hard at them, and as a result have learned things that they are not likely to forget. The games and activities presented in this chapter represent a natural extension of that most successful learning method.

We begin this chapter on early math concepts with four introductory activities not requiring the use of numerals: "My Own Counters," "Size-Place Strips," "The Counting Board" and a counting game for four players—"Going to the Zoo." Numerals are then introduced in such games as "Mini-Bingo" and "Hop-Skip-Jump;" in individual activities such as "Train Set," and in the whole class activity of "Number Clothesline."

Students might enjoy closing their eyes and trying to identify some "Sandpaper Numbers" or sorting a large quantity of beans in "Egg Carton Counting." "Calendar Bingo" uses a few old calendars and converts them into a game involving recognition of numbers up to 31.

"Bean Strips" is a game which serves as a model for place value. The materials made for this game may be used for "Bean Strip Trading," another game in this chapter, and for two later games in the next chapter on addition and subtraction.

"Compare" is an activity for the whole class involving the concepts of "greater than" and "less than," while small groups of students will enjoy choosing cards on the basis of place value as they travel through "Hundreds Land"—a board game for two to four students. Finally, students may go around the "Rounding Ring" only if they can round numbers to the nearest tens, hundreds, and thousands.

By selecting the games and activities which fit the needs of your classroom and coordinating them with your mathematics curriculum, your students will most certainly be learning through play.

MY OWN COUNTERS

Summary: Students are given their own counters which can be used for various concrete experiences with numbers. Counters are stored in individually decorated containers. A whole class activity.

Skills: Counting

Supplies: 10-20 counters per student (chips, buttons, beans, or small cubes), containers for the counters (students should bring in small empty boxes), crayons, construction paper, paste.

Preparation: Each student will need 10 counters and a container for them. Students write their names on their containers and decorate them, each in their own individual way.

Directions:

Students can use these counters throughout the school year. Later, increase the number of counters to 20.

Counters should be available for students to use for counting, for one-to-one correspondences, for discussions of "one more" and "one less," and for simple addition and subtraction problems.

Allow students to make up their own games with counters. For example, a student might hold a number of counters in his or her hand, while other students try to guess "how many."

Subtraction could be approached intuitively through this little game: A student places a number of counters on a desk, which should be counted. Then student closes his or her eyes. Another student takes away a few of these counters. After opening eyes, student must figure out how many counters were removed.

SIZE-PLACE STRIPS

Summary: Students are each given five strips which they cut out, compare in length, and arrange in order. A whole class activity.

Skills: Comparing lengths, ordering

Supplies: Paper for duplicating, pen, scissors, and paste (optional).

Preparation: Draw and reproduce Figure 1-1(a). Give one sheet to each student.

Directions:

1. Each student should cut out a set of five strips, each one a different length.

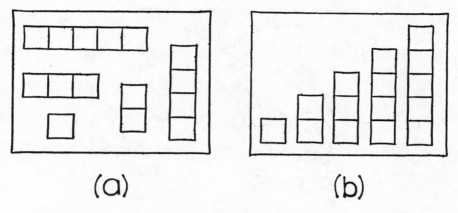

(a)   (b)

Figure 1-1

2. Students arrange the strips in size places from the one-square to the five-square length, as in Figure 1-1(b).

3. If desired, students could then paste the strips onto blank sheets of paper.

4. You might wish to compare the strips to the students' ages— how they grew from little one-year-olds to two-, three-, four- and five-year-olds, and how they kept getting bigger and bigger.

THE COUNTING BOARD

Summary: Students come up to the counting board to answer questions involving counting. A whole class activity.

Skills: Counting

Supplies: Tagboard, pens of various colors.

Preparation: To make the counting board, draw different objects, in different colors with slight variations (as in Figure 1-2) on a large piece of tagboard. For example, have some blocks blue, others red, some brown bears and some orange ones, yellow stars and blue stars.

Directions:

Hang counting board on classroom wall. Allow students to come up to the board to answer questions:

How many alphabet blocks?

How many red blocks?

Figure 1-2

How many "A" blocks?

How many teddy bears?

How many teddy bears with bows?

Later on, let students make up questions.

GOING TO THE ZOO—A COUNTING GAME

Summary: A game for two to four players. Students move their markers around a gameboard by counting objects. No numerals are used.

Skills: Counting

Supplies: Tagboard, pen, four game markers (different colors), nine beans or other counters, small amount of paint, small empty can.

Preparation: Using a large square of tagboard, prepare a gameboard similar to the one shown (Figure 1-3). Take nine beans or some other counters and paint each on one side only.

Directions:

1. Each player places his or her marker on square marked "START."

2. In turn, each player places the nine beans in the empty can, shakes it, and rolls the beans out.

3. Player counts the number of beans that have landed with their painted sides up, and moves marker that number of squares.

4. Optional: If marker should land in a square with picture of balloons, then player counts the balloons and moves marker that number of spaces extra.

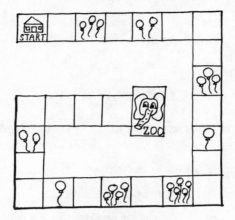

Figure 1-3

5. First player to reach the zoo is the "first winner," but allow all children to finish and be "winners." (It is not necessary to reach the zoo by the exact number.)

TRAIN SET

Summary: Students arrange five (or more) cut-out trains in numerical order. Self-correcting.

Skills: Ordering numbers

Supplies: Tagboard, felt-tip pens.

Preparation: Draw and cut out a series of trains representing the numbers from 1 to 5 (or higher). Each train has a numeral written on it and a corresponding number of windows drawn on it.
Optional: Make the trains with connectors cut out in different shapes so that when trains are connected, numbers will be in correct order (Figure 1-4).

Figure 1-4

Directions:

Working independently or in small groups, students arrange trains in numerical order.

HOP-SKIP-JUMP

Summary: Students do a series of activities, each a different number of times, according to numerals shown. A whole class activity.

Skills: Recognizing numerals, counting

Supplies: Construction paper, pen, scissors.

Preparation: Cut construction paper into large rectangles—20 will be needed. On 10 of the rectangles, write the digits 0 to 9. The other 10 cards will have activities such as, clap, hop, skip, jump, bend over, shake hands, spin around, fall down, crawl, and stretch up. Each activity card should have both the word, such as "JUMP," and a stick-figure illustration.

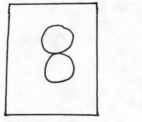 

Figure 1-5

Directions:

1. Two students come up to the front of the classroom. One is given the numeral cards, the other the activity cards.

2. At the same time, each holds up one card.

3. The rest of the students are to do the activity the number of times shown. "CLAP" and "4" means the class should clap hands four times.

4. This is repeated until all cards have been used.

INDIVIDUAL NUMBER PUZZLES

Summary: A self-correcting activity where students match numbers to corresponding pictures.

Skills: Matching numbers

Supplies: Tagboard, pen, scissors.

Preparation: Cut tagboard into 10 rectangles. Write numerals on the top half and draw pictures representing the numbers on the bottom half. Cut each rectangle in a unique way, as shown in Figure 1-6.

Directions:

Give students the puzzle parts and show them how pieces will fit together when number is correctly matched with picture.

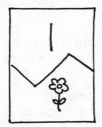   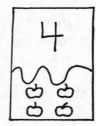

Figure 1-6

MINI-BINGO

Summary: Students cover numerals on a bingo card, hoping to be first one to cover all nine squares. A whole class activity.

Skills: Recognizing numerals

Supplies: Construction paper, pen, markers (nine per student).

Preparation: For each student draw a 3x3 array on a sheet of construction paper. Choose nine numerals from 0 to 10 and write them in the nine squares in random order. Use different groups of nine numerals and make as many varied arrangements of the numerals as possible, so that each card will be different. Note that each card will be missing two numerals (Figure 1-7).
Cut and label 11 cards with the numerals 0 to 10.

Directions:

1. Each student should be given a bingo card and nine markers.
2. Teacher or a chosen student mixes the 11 cards, then chooses one and announces the number. If necessary, show number to the class.

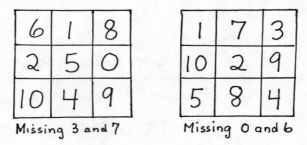

Missing 3 and 7   Missing 0 and 6

Figure 1-7

3. All students who have that number on their card cover it with a marker.

4. More numbers should be chosen until a student has covered all nine squares on his or her card. (There may be more than one winner.)

5. In later games, students can try to be first to cover an "X" shape or an "L" shape, a "T" shape or a "U" shape on their bingo cards.

NUMBER CLOTHESLINE

Summary: Students move cut-outs of clothing on a clothesline until all pieces are in numerical order. A whole class activity.

Skills: Ordering numbers

Supplies: Construction paper, pen, string (clothespins optional).

Preparation: Fold construction paper and draw outline of clothing. Cut both sides back to back, connected only on the top. When open, cut-outs should resemble drawing below. (Figure 1-8). Label with numerals on the outside.

Directions:

1. Have two students hold up clothesline.
2. Folded clothing is placed on line in any order. Clothespins are not really needed but may be used for fun.
3. Students come up one at a time and move clothing so that the piece labeled "1" is first, piece labeled "2" is second, and so forth until all pieces are in correct order.

Variation: Place addition or subtraction problems on the outside of the clothes and answers on the inside. Student picks a piece of clothing from the clothesline, answers question, then checks answer inside.

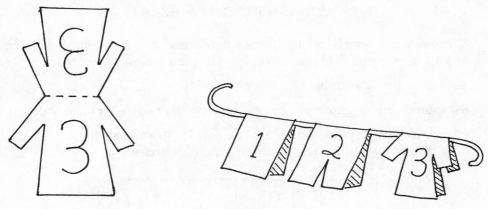

Figure 1-8

EGG CARTON COUNTING

Summary: Beans are placed into sections of an egg carton according to the number in the section.

Skills: Recognizing numerals and counting

Supplies: Egg carton, pen, beans or other small counters.

Preparation: Remove cover from an egg carton and label sections with different numerals.

Directions:

1. Give student an egg carton and the correct number of beans. A carton labeled as in Figure 1-9 would need 30 beans, while a carton labeled 1 to 12 would need 78 beans.

2. Have students place no beans in section marked "0," one bean in section marked "1," two beans in section marked "2," until all sections are filled appropriately.

3. If done correctly there will be no remaining beans.

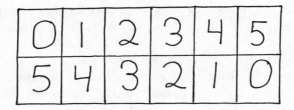

Figure 1-9

CUT AND PASTE NUMBERS

Summary: Numerals not in correct numerical order originally are cut apart, rearranged, and pasted down. A whole class activity.

Skills: Ordering numbers

Supplies: Construction paper, pen, scissors and paste for each student.

Preparation: Prepare a strip for each child, 10 squares long, with the numerals 1 to 10 written in an incorrect order.

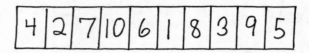

Figure 1-10

Directions:

1. Each student should be given a number strip, a pair of scissors, a sheet of construction paper, and paste.
2. Students are to cut out the 10 number squares, arrange them in the correct order, and paste them onto a sheet of construction paper.

NUMBER BOOKS, NUMBER CHARTS AND NUMBER CARDS

Summary: The teacher writes the numerals and the students draw pictures or paste objects to illustrate each number.

Skills: Recognizing numerals and counting

Supplies: Some of the following supplies can be used: construction paper, tagboard, crayons, paste, small objects (beans, macaroni, buttons).

Directions:

Students might like to make number books, number charts, or number cards, which match numerals with the appropriate number of objects.

For books, the teacher should write numerals 1-5 (or 1-10) on separate pages; then have students draw pictures under the numerals. For example: 1 tree, 2 cars, 3 balls, 4 houses, 5 flowers. When finished, a special cover should be drawn, and pages should be attached together.

For number cards or number charts, instead of drawing pictures, students might paste a variety of objects like beans, macaroni or buttons onto 10 cards numbered from 1-10, or one large chart containing all 10 numbers.

SANDPAPER NUMBERS

Summary: Students feel and identify sandpaper numerals. For a small group.

Skills: Recognizing numerals, ordering numbers

Supplies: Sandpaper, nine blank cards, and white glue.

Preparation: Cut the numerals 1 to 9 from a sheet of sandpaper and glue onto nine cards.

Directions:

The following are some of the activities that might be done with sandpaper numbers:

1. Allow the students to trace the shape of each number with their fingers as they identify the name of the numeral.
2. Have students arrange the sandpaper numbers in numerical order.
3. Have students close their eyes. Let each student identify a numeral by feeling it.

MONEY MATCH

Summary: Students match pictures and prices of items for sale with cards showing the exact amount of money. Can be played as solitaire or as a competitive game for four players.

Skills: Understanding money—pennies, nickels and dimes

Supplies: 24 blank cards (index cards or tagboard), felt-tip pen.

Preparation: On 12 cards draw pictures of simple objects that students might buy. Pictures may be cut from magazines. Include the name of object and its price (Figure 1-11). On 12 other cards show the amount of money needed to buy each item; use 1¢, 5¢ and 10¢.

Directions:

For Solitaire: Student spreads the 12 picture cards out on a table. Each money card is placed on top of correct item with the matching price until all cards are matched.

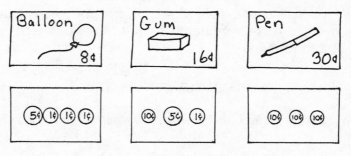

Figure 1-11

For Four Players:

1. The 24 cards are mixed and six cards are given to each player.

2. Students look at their cards for a matching pair (picture plus money) and place such a pair face up in front of them. This represents a purchase. The other students should check to see if the price is correct.

3. Then, students each take one of their remaining cards and, at the same time, all students pass a card to the student on the right. Students look for a new match, which would again be placed face up and checked.

4. This continues until one player has made three matches.

CALENDAR BINGO

Summary: This game, for the whole class, is made up of old calendars and played like bingo.

Skills: Recognizing numerals from 1 to 31

Supplies: Old or unused calendars (about three), pen, markers (12 per student).

Preparation: Using old calendars, cut a 4x4 square (containing 16 numerals) out of each page. Cut one for each student. To make each sheet unique, cross out (with an "X") two different squares on each page (Figure 1-12).
Find one month that has 31 days and cut it up into 31 squares; this gives you the numerals 1 to 31.

Directions:

1. Give each student a calendar bingo sheet and 12 markers.

Figure 1-12

Students immediately place markers over the two free squares containing Xs.

2. Game is now played like bingo. The cut-up calendar numerals are picked one at a time, called out and/or written on the blackboard. Meanwhile, students cover these numerals with markers.

3. The first student to cover four in a row is the winner.

BEAN STRIPS

Summary: Students glue beans to strips of tagboard or popsicle sticks to help prepare this game, which serves as a model for place value. The materials here may also be used for two games in the next chapter, "The 100-Bean Game" and "99-Bean Subtraction."

Skills: Place value

Supplies: Tagboard, package of dry beans, white glue, pen. (Tongue depressors or popsicle sticks can be used instead of tagboard strips.)

Preparation: Cut tagboard into long strips 1 inch by 6 inches (about 2cm x 16cm) and glue 10 beans to each strip (Figure 1-13). Or, glue beans to popsicle sticks. If possible, have each student make two bean strips. For each group of five students, make a playing board

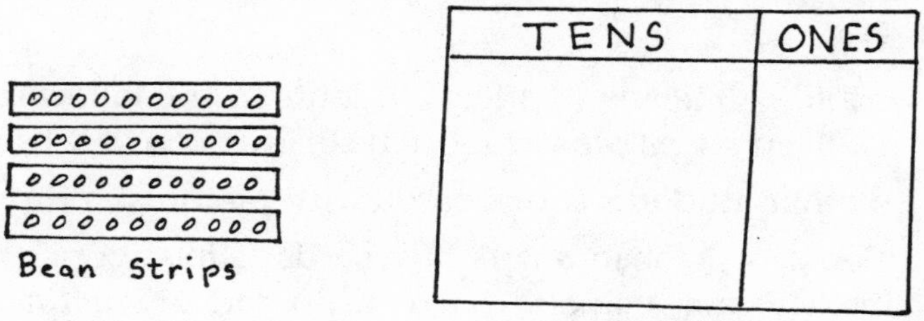

Bean Strips

Figure 1-13

consisting of two columns: one labeled "TENS," the other, "ONES." Also have some loose beans available.

Directions:

1. Divide class up into teams, with five students on each team.
2. Each team should have 10 bean strips, 10 loose beans, and one playing board.
3. Teacher calls out and/or writes a two-digit number on the board.
4. Each team tries to place the proper number of bean strips and single beans on playing board (in correct columns) to illustrate that number.
5. When all members of the team agree that the bean strips are correct, they raise their hands.
6. Teacher gives two points to the first team that is correct and one point to all other teams also correct. After 10 rounds, team with the most points wins.

Note: These bean strips can later be used to illustrate three-digit numbers. To form hundreds, wrap groups of 10 bean strips with rubber bands.

BEAN STRIP TRADING

Summary: Using bean strips from previous game, students buy and sell objects.

Skills: Place value

Supplies: Bean strips (Figure 1-13) and loose beans, containers for bean strips, and index cards with pictures and prices of simple objects that students might buy (Figure 1-11). Prices should be no higher than 50¢, with some cards less than 20¢.

Directions:

1. Distribute 50 beans to a few students in the following manner: give 4 bean strips plus 10 loose beans to each of these students.
2. Give other students index cards with pictures and prices.
3. Students with bean strips buy cards using proper amount of beans. This may be done in a highly organized manner, with one student trading at a time; or it may be done more freely with students walking around the room buying and selling.

4. As students receive beans for payment, they may, in turn, use beans to buy other cards.

5. After all possible trades have been made, materials may be collected and redistributed with different students starting out with the bean strips.

6. *Optional:* Appoint one student to be banker. This student is given about 50 loose beans. Banker will trade 10 loose beans for a bean strip when needed. *Example:* a student has three bean strips and two loose beans and wishes to buy an item that costs 27¢.

COMPARE

Summary: In this game, two-digit numbers are compared between a "Greater Than" team and a "Less Than" team. (The game may be adapted for one-digit or three-digit numbers.)

Skills: Comparison of numbers

Supplies: 22 large index cards or squares of tagboard; pen.

Preparation: On one large card write the symbol < (less than) and on another card write the symbol > (greater than). On the remaining 20 cards write the numerals: 12, 16, 20, 25, 30, 31, 39, 43, 44, 48, 52, 56, 57, 61, 69, 70, 75, 80, 84, 93.

Directions:

1. Shuffle the 20 number cards and place in a face-down pile on a desk in front of the room. Two student volunteers should come to the front of room.

2. Divide class into two teams. The left side of room will become the "Less Than" team, while the right side becomes the "Greater Than" team.

3. The student volunteer who is standing on the left (from the class's perspective) will draw the top card and hold it up. The other student volunteer will take the next card and hold it up.

4. The class must decide on the correct symbol to be placed between the two numbers.

5. If it is a "greater than" symbol, a student from the "Greater Than" team will come up and place > between the two numbers. If it had been <, it would be placed by a student from the "Less Than" team.

6. The winning team will read the correct number sentence (for example, "61 is greater than 48.") and that team wins one point.

7. Game continues and after 10 pairs of numbers have been selected, the team having the most points wins the game.

HUNDREDS LAND

Summary: For two to four players. Students choose cards and move across a gameboard according to the digits corresponding to the different sections of the board: hundreds, tens and units.

Skills: Place value

Supplies: Oaktag, 10 index cards, felt-tip pens (three colors), four different markers.

Preparation: Prepare a gameboard (Figure 1-14), using different colors for the three sections. Cut the index cards in half, making 20 small cards. Label each of these 20 cards with a different three-digit number.

Directions:

1. Dealer gives each player five cards.

2. Each player places marker on "START." Movement in the game depends upon which section of the gameboard the player's marker is currently in.

3. Players, in turn, first select one of their cards and place it face up on "card pile" in the center of gameboard.

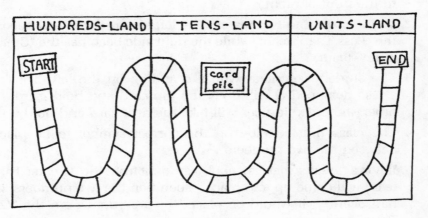

Figure 1-14

4. If player's marker is in Hundreds-Land (as it must be at the start of the game), player will move the number of spaces corresponding to the hundred's digit of the selected card. (For example, if card reads 672, move six spaces.)

   When player is in Tens-Land at the beginning of a turn, player moves according to the ten's digit. (If card is 489, move eight spaces.) Likewise, when player is in Units-Land, movement corresponds to the unit's digit.

5. Player always begins by first selecting a new card and then moving marker. After all cards have been discarded, dealer mixes up the cards and again gives each player five cards.

6. Game continues and first player to reach "END" wins.

ROUNDING RING

Summary: A board game for two to four players, where students round numbers to nearest tens, hundreds, and thousands.

Skills: Rounding numbers

Supplies: Tagboard, four different markers, one die (dice), self-sticking adhesive labels or tape, pen.

Preparation: Prepare a gameboard similar to the one shown in Figure 1-15. Prepare dice by placing labels on six sides of the die and numbering sides 1,2,3 (each twice). See *Note* on preparing Answer Key.

Directions:

1. Students place markers on "START," and in turn, throw the die. The number will be used to tell students how many spaces to move, and also how to round the number.

2. If a 1 is thrown—student moves one space and rounds number landed on to the nearest tens.

   If a 2 is thrown—student moves two spaces and rounds number landed on to the nearest hundreds.

   If a 3 is thrown—student moves three spaces and rounds number landed on to the nearest thousands.

3. If correct, student stays in that space. If not correct, student moves back to previous position.

4. First student to go around the ring is the first winner, next student to complete the ring is the second winner, etc.

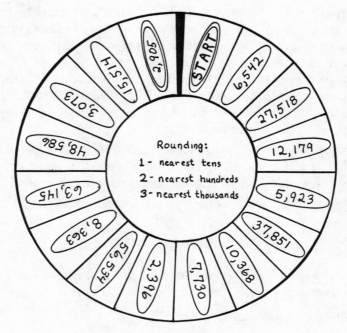

Figure 1-15

Note: An Answer Key may be made up, kept face down, and only used if an answer is challenged. Answer Key would begin like this:

ORIGINAL NUMBER	NEAREST TENS	NEAREST HUNDREDS	NEAREST THOUSANDS
6,542	6,540	6,500	7,000
27,618	27,620	27,600	28,000
...	...	...	...

CHAPTER **2**

Making Effective Games
That Teach and Strengthen
Addition and Subtraction Skills

Just as important as the building of basic addition and subtraction skills, is the development of positive attitudes toward mathematics. The materials presented in this book have been developed to meet both these objectives.

While most of the games in this chapter need some preparation, they are meant to become permanent additions to your classroom. To offer your students a variety of approaches to addition and subtraction, combine the appropriate games in this chapter with those of Chapters 5 and 6, which require little or no preparation. Be sure to supplement these activities with addition puzzles, simple logic games, geometry projects and the addition card games found in later chapters.

In this chapter, "Top Card" is an excellent game for reinforcing addition combinations that can be adjusted to fit your students' individual abilities. "Tic-Tac-Ten" can be played on different levels with beginning students involved only in the additions, while more advanced students will start developing game strategies. Both "Let's Go Fishing" and "Picture Puzzle" are self-correcting, while the game of "Indian Stones" allows the students to ask the questions. Subtraction skills can be strengthened by two interesting games— "Spanish Subtraction" and "Subtraction Spin." A humorous change of pace will be provided by "Silly Stunts," as students are encouraged to learn more difficult addition facts.

The "100-Bean Game" and "99-Bean Subtraction" offer a model for addition and subtraction with regrouping. However, both can be played before regrouping has been introduced formally. This is followed by a few games that require addition and subtraction of two- and three-digit numbers: "Post Office," "Beehive," and "Supermarket."

A game that students are sure to enjoy is "Pac-Math," a variation of the video game, *Pac-Man*, where players add or subtract two-digit numbers (with regrouping) to control the little figures as they run around the gameboard.

MATH COLORING

Summary: Answers to arithmetic problems determine which colors to use. For students already familiar with this type of activity, a variation is suggested.

Skills: Addition or subtraction

Supplies: Pictures from coloring books, pencil, crayons.

Preparation: Take uncolored picture from coloring or design book. Make a color key. Place appropriate addition or subtraction problems in areas to be colored, as in Figure 2-1. Extra lines might be drawn in. Duplicate for each student. (For the variation, do not place addition combinations onto the picture.)

Directions:

Each addition or subtraction example is to be worked out and an area is to be colored correctly according to color key. If necessary, allow students to use counters to solve problems.

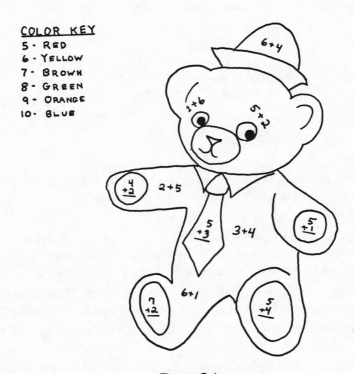

Figure 2-1

Variation: Give students uncolored pictures. Set up a color key. Using this key, students make up their own addition and subtraction questions and put them on the picture. Later, students exchange papers and color in each other's pictures.

TOP CARD

Summary: Using prepared cards with addition combinations, this game parallels the familiar card game "War." As students progress, new combinations can be added and some of the originals removed.

Skills: Addition

Supplies: Tagboard or index cards, pen, scissors.

Preparation: Cut 48 cards out of tagboard or use 24 index cards and cut each in half. Label cards with basic addition combinations. Always have at least two cards for each sum. Reverse the order on second card, 2+4 would become 4+2. Begin with only a few combinations repeated many times. Introduce new numbers only after the students have mastered the original combinations.

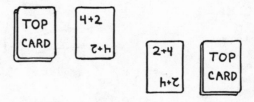

Figure 2-2

Directions:

1. Divide cards evenly among the players. Game is usually played with two students but three or four may play.

2. Players stack cards face down.

3. Each student turns over top card and places it on table face up. Students should call out their sums. In the beginning, students could be allowed to use counters to work out sums.

4. Player with the highest sum wins the cards. These cards are now placed face down at the bottom of this winning player's pile.

5. Game continues with the players again turning over their top cards.

6. If two or more players have the same highest sum, then to break the tie, they place the next three cards face down on the table and a fourth card face up. The player whose fourth card has the

highest sum wins all these cards. (If any of the players had less than four cards left, then they would turn over their last card.)

7. Winner is the player who accumulates all the cards. Or, if a time limit is desired, player having the most cards after a given amount of time wins.

Variations: Cards may be labeled with subtraction, multiplication or division questions.

LET'S GO FISHING

Summary: Fish are caught by gathering all fish with a specific sum.

Skills: Addition (sums to 10)

Supplies: Tagboard, felt-tip pen.

Preparation: Using tagboard, cut out 35 fish-shaped pieces. Also cut seven long rectangles to represent fishing poles. (See Figure 2-3.) Label each fish with a different combination: $0+4, 1+3, 2+2, 3+1, 4+0; 1+4, 2+3, 3+2, 4+1, 5+0; 1+5, 2+4, 3+3, 4+2, 5+1; 1+6, 2+5, 3+4, 4+3, 5+2; 2+6, 3+5, 4+4, 5+3, 6+2; 3+6, 4+5, 5+4, 6+3, 7+2; 4+6, 5+5, 6+4, 7+3, 8+2.$ Label the seven fishing poles: 4,5,6,7,8,9,10. In this way there will be five different combinations for each sum.

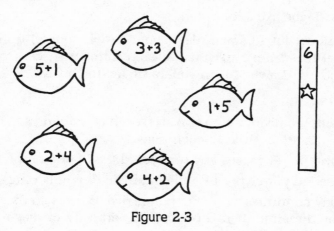

Figure 2-3

To make the game self-correcting place a different symbol, such as a blue star or a red square, on each fishing pole. Then place the appropriate symbol on the back of each fish with the same sum. If the "6" fishing pole has a blue star, then a blue star will be placed on the backs of all fish having a sum of 6.

Directions:

1. Place all fish on floor or table so that addition combinations can be seen.

2. Pick a student and give him or her a fishing pole. The object of this game is to have the student find all five of the fish that have a sum equal to the number on his or her fishing pole.

3. After each fish is picked, student needs only to look at the back of the fish to see if this symbol matches the one on the fishing pole. Other students may help.

4. After all five combinations are found, another student is selected and given a different fishing pole. Continue until all the poles are given out, and all the fish caught.

Variation: Use subtraction combinations.

TIC-TAC-TEN

Summary: The object is to place the third number on a tic-tac-toe board, so that the sum of the three numbers is equal to 10. While the game appears simple, it can be played on a more complex level.

Skills: Addition, subtraction, logical thinking

Supplies: Tagboard, felt-tip pen.

Preparation: Cut 10 small tiles (about 1-inch square) from tagboard. Label tiles from 1 to 5, using each number twice. On a piece of tagboard, draw a 4-inch square tic-tac-toe board.

Directions:

1. Tiles are divided into two sets, each containing a 1, 2, 3, 4, and 5 tile. A set is given to each player.

2. First player places any numbered tile in one of the nine boxes.

3. Second player places a tile in any of the remaining eight boxes.

4. Play continues until one player places a tile so that the sum of the three numbers (horizontally, vertically, or diagonally) is ten, thereby winning the game.

5. Players alternate going first. At the end of ten games, the student who won the most games is the winner. If players are evenly matched it may be a tie. Also, individual games can end up in a draw with no winner, as in Figure 2-5.

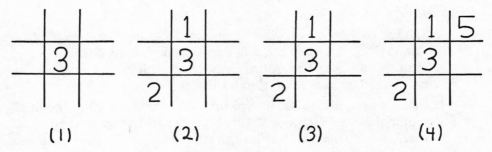

Figure 2-4

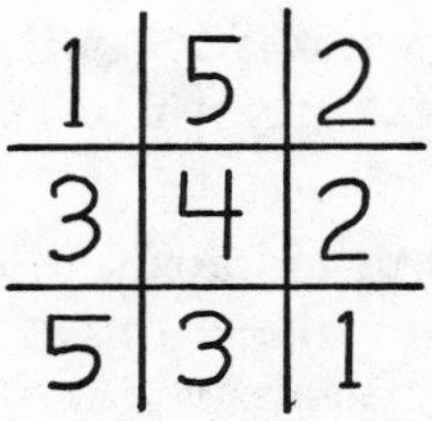

$$\begin{array}{|c|c|c|} \hline 1 & 5 & 2 \\ \hline 3 & 4 & 2 \\ \hline 5 & 3 & 1 \\ \hline \end{array}$$

Figure 2-5

Note: Some moves are strategically better then others. In the sample game (Figure 2-4), the first player placed a 3 tile in the center box; the second player put a 1 tile down. Had the second player put down a 2 instead, the first player could have won immediately with a 5 tile. But the first player cannot win; since $3+1=4$ and $10-4=6$, a 6 tile would be needed for a sum of 10 and, of course, there is no 6 tile.

Another good second move in these circumstances would be a 4 tile. Again, the first player could not win. Since $3+4=7$ and $10-7=3$, a 3 tile would be needed to win, but the first player has already used it. It is useful to know what tiles the other player has left.

PICTURE PUZZLES

Summary: By matching addition or subtraction problems with their answers, students put together picture puzzles. Self-correcting.

Skills: Addition or subtraction

Supplies: Two page-size sheets of cardboard or tagboard for each puzzle; full-page pictures cut from magazines or books; pen and white glue; scissors.

Preparation:

1. To glue picture on cardboard, first apply glue to back of picture. Then, using an extra strip of cardboard, spread a thin layer of glue evenly over the entire back of the picture. Place on the cardboard, smooth out, and let dry.

2. Using a pen, divide cardboard on back of picture into nine or 12 rectangles. See Figure 2-6. Do the same to the second piece of cardboard. This will be the question board. On each of these rectangles write an addition or subtraction problem. These can be easy as 3+2, which students can do in their heads or with counters, or more difficult ones, such as 64−28 where students may need scrap paper. It is important that each problem have a different answer. You cannot use 12−4 and 10− 2 in the same puzzle.

3. On the rectangles on the back of the picture write corresponding answers. However, these must be arranged as in a mirror image. For example, the answer to the lower left rectangle on the question board should appear on the lower right corner on the back of puzzle. This is because the picture pieces will be turned over before being placed on top of the board.

4. Cut the picture puzzle up into rectangular pieces. Leave the question board whole.

5. If more than one puzzle is to be made, use a different color for each puzzle.

Directions:

1. Place question board on table. Take pieces of the picture puzzle and place them all with the answers on the back facing up.

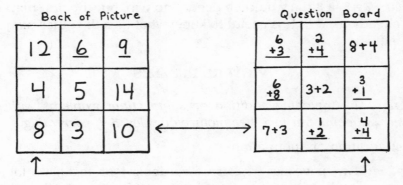

Figure 2-6

Figure 2-7

2. As each question is worked out, each piece believed to be the correct answer is turned over, picture side up, and placed over the question. See Figure 2-7.

3. The puzzle will be self-correcting as the picture begins to take shape. Correct answers will form a perfect picture.

Variations: Multiplication or division.

SILLY STUNTS

Summary: Students answer addition questions and then perform silly stunts as directed on back of answer card.

Skills: Addition (sums to 18)

Supplies: 24 index cards, pen, scissors.

Preparation: Cut the index cards in half, making 48 halves. 24 will be for addition combinations and 24 will be for the sums. Place these combinations on 24 card-halves: 3 + 8, 7 + 4, 5 + 6, 9 + 2; 8 + 4, 5 + 7, 6 + 6, 3 + 9; 4 + 9, 7 + 6, 5 + 8, 6 + 7; 8 + 6, 7 + 7, 5 + 9; 9 + 6, 7 + 8, 8 + 7; 8 + 8, 7 + 9, 9 + 7; 9 + 8, 8 + 9; 9 + 9.

On remaining 24 card-halves write the numbers 11 through 18, three times each. These represent the sums. Turn the sum cards over and write different "silly stunts" on each. Here are some examples, but you, as the teacher, would be the best judge as to what silly stunts are appropriate for your class. Perhaps students could assist in thinking of some.

"Hop on right foot 12 times"

"Walk around classroom with eyes closed"

"Spin around 7 times"

"Walk across front of room, balancing a book on head"

"Whistle 5 times, then sing a song"

Directions:

1. Place all 24 cards with addition combinations face down in one pile. This is the question pile.

2. Divide the 24 cards with the sums into eight piles according to number on cards. There should be three cards in each pile. These are the answer piles. Have numbers face up, and stunts face down as in Figure 2-8.

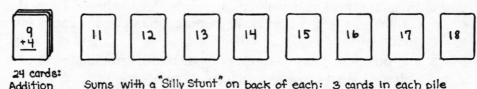

24 cards:
Addition
Combinations

Sums with a "Silly Stunt" on back of each: 3 cards in each pile

Figure 2-8

3. Student takes a card from the question pile, reads it to the class and then announces the probable sum. Class decides if answer is correct; if not, they give the correct answer.

4. In either case, the student selects the top card from the answer pile having that correct sum on it. Student then performs this stunt.

5. Have student place the answer card back on the bottom of the pile it came from. There should always be three cards in each answer pile.

6. This student now selects another student to pick a question card and the game continues.

INDIAN STONES

Summary: When a team lands on a numbered stone, the opposing team makes up an addition question to challenge this team.

Skills: Addition (sums to 20)

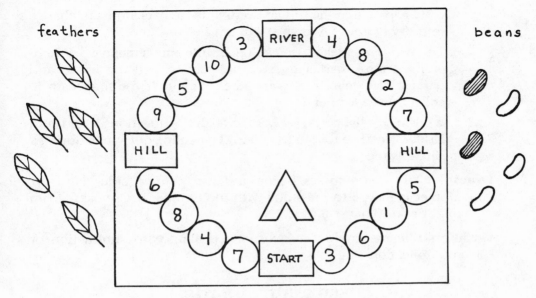

feathers   beans

Figure 2-9

Supplies: Tagboard, felt-tip pens, two different markers, and five beans painted on one side.

Preparation: Draw a gameboard similar to one in Figure 2-9. Use a few colors to make it attractive. The five beans should be painted on one side only. From the tagboard, cut five small feather-shaped pieces.

Directions:

1. Players divide up into two teams. Play by tossing the beans and moving marker the number of painted sides that land up. Teams take turns.

2. If the marker lands on a stone with a number, the team must pass an "Indian Brave Test." An addition question based on the number landed on will be made up by the opposing team. It should be a fact already discussed in class, and the team asking the question should know the answer.

3. If the question is answered correctly, marker will stay on stone. If a mistake is made and asking team knows the correct answer, then as a penalty, marker must be moved back five spaces.

4. If marker lands on Hill, this team loses a turn. (The idea here is that it takes extra time to climb up the hill.)

5. If marker lands in the River, it must be returned to start. (Here we

will imagine that the Indian Brave must go back home to change into dry clothes.) A feather will not be collected.

6. The two teams take turns going around the gameboard. Whenever a team's marker goes completely around the board and passes start, team earns a feather. Winner is the first team to collect three feathers.

7. If desired, a list could be written of all the questions asked, and students can be told to try to avoid repeating questions during a single game.

Example: First team tosses beans and two painted sides land up. Marker is moved to the stone with the number 6. Opposing team might ask, "What is 6 plus 7?"

Variations: Teams could ask questions dealing with subtraction or multiplication.

EGG CARTON ADDITION

Summary: Players shake egg carton and add three or more of the numbers inside.

Skills: Addition of three or more numbers

Supplies: An egg carton with solid cover; three or more small objects (beans or buttons); paper and pencil for each player.

Preparation: Write a different number in each of the 12 sections of the egg carton.

Directions:

1. Decide how many numbers are to be added. Use the same number of small markers.

2. Each player, in turn, places markers in egg carton, closes cover and shakes carton horizontally.

3. Egg carton is placed on table and cover opened.

4. Score is found by adding up all the numbers in sections where the markers landed. If two markers land in the same section, add that number twice. If three markers are used, three numbers should be added. If four markers, then add four numbers, etc.

5. The players should add each other's scores, checking for errors. Highest score wins.

SPANISH SUBTRACTION

Summary: Players throw numbered dice, compute the difference between the numbers, and attempt to clear their section of the board of all playing pieces.

Skills: Subtraction

Supplies: Tagboard, two dice, self-adhesive labels, pen, 24 beans or other small objects.

Preparation: Make a gameboard similar to the one in Figure 2-10, with the rules written on the gameboard. Next, prepare the dice (also used in the game "Take a Trip"). Using self-adhesive labels, write the numerals 1 to 6 for the first die, and 5 to 10 for the second. Cut to size and stick on the dice.

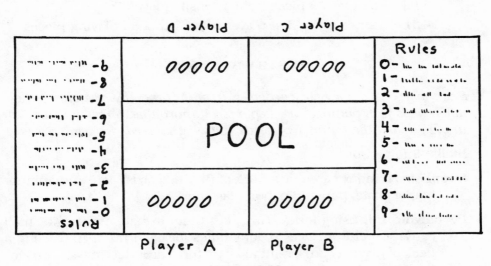

Figure 2-10

Directions: (For four players)

1. Each player gets five beans to be used as playing pieces. These are placed on a section of the board. The remaining four are placed in the Pool area.

2. First player throws dice, and subtracts smaller number from larger. This result is then looked up on the rules chart and player follows directions:

SUBTRACTION RESULT	RULES FOR PLAYING PIECES
0 —	Throw dice again
1 —	Give 1 to Opposite Player
2 —	Take 1 from Pool
3 —	Take 1 from Opposite Player
4 —	Give 1 to Pool
5 —	Give 2 to Pool
6 —	Give 1 to any other Player
7 —	Take 1 from any other Player
8 —	Give 1 to Pool and Go again
9 —	Take 1 from Pool and Go again

3. An example: Player A throws 5 and 8 on dice, computes that $8 - 5 = 3$, and looks up 3 on the rules chart. Player A would then take one playing piece from opposite player, D.

4. The winner is the first player to get rid of all playing pieces.

SUBTRACTION SPIN

Summary: Cards are discarded two at a time, by matching their differences with numbers on a spinner. If more than four students wish to play, adjust the number of cards, allowing 10 cards for each player.

Skills: Subtraction

Supplies: 20 index cards or a deck of playing cards, pen, and spinner (paper plate, paper clip, brass fastener).

Preparation: If using index cards, cut them in half, making 40 small cards. Label from 1 to 10, four times for each number. If using a deck of playing cards, remove all picture cards, leaving 40 cards for the game. In using playing cards, the ace represents 1 and all others represent their face value.

Prepare spinner as in Figure 2-11. First the paper plate is to be divided into 10 sections, labeled 0 to 9. Then attach a brass fastener

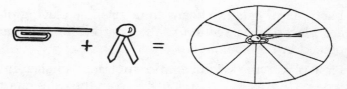

Figure 2-11

through an opened paper clip and then through the center of plate—loosely enough to allow the paper clip to spin freely.

Directions:

1. Mix cards and give 10 cards to each player. They are to be placed face up in front of each player on a table or on the floor.

2. The idea is to get rid of cards, two at a time.

3. A number is spinned, say it is 3. Each player looks at his or her cards. Do any two cards have a difference of 3? If a pair is found, such as 8 and 5, it is discarded. For a single spin each player can discard only one pair.

4. Another player spins; spinner lands on a different number. All players look for a pair of cards which have that number as the difference.

5. First player to get rid of all cards wins. Game can continue until other players discard all their cards.

TAKE A TRIP

Summary: The players throw prepared dice and move according to the sum or difference of the two dice.

Skills: Addition and subtraction

Supplies: Tagboard, two dice, self-adhesive labels, felt-tip pens, four different markers.

Preparation: Using tagboard, make a gameboard similar to the one in Figure 2-12. To prepare the dice (also used for "Spanish Subtraction") use self-adhesive labels and write the numerals from 1 to 6 for the first die, 5 to 10 for the second. Cut to size and stick on dice.

Directions:

1. Players (up to four) each place a marker on "HOME." Object of the game is to be the first to reach "END OF TRIP."

2. First player throws dice and can move his or her marker according to either the sum or the difference of the numbers shown on dice. The decision will be guided by the following rules:

   a. If marker lands on an X-space, lose a turn.
   b. If marker lands on an O-space, this is an airport; fly to connecting airport if it is in the right direction.

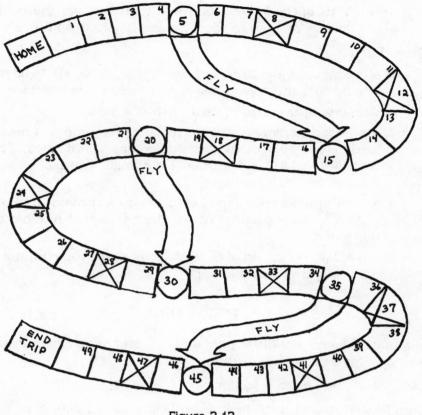

Figure 2-12

   c. If marker lands on another player's square, then this player must move back five spaces.

Example: Player is on "HOME," and throws a 3 and an 8 on the dice. Player may move five spaces or 11 spaces. In this case, it would be to the player's advantage to move to square 5, which is an airport.

THE 100-BEAN GAME

Summary: This game can be played before addition with exchanges has been introduced formally. Later, it can serve as a model. In the game, two teams (or two students) collect beans, trade loose beans for bean strips whenever possible, and try to be the first to accumulate 100 beans.

Skills: Addition—model for addition with exchanging (usually called carrying, regrouping, or renaming)

Supplies: Tagboard, package of beans (at least 240), white glue, two dice, pen. (Materials also may be used for "99-Bean Subtraction.") Popsicle sticks can be used instead of tagboard strips.

Preparation: Cut part of the tagboard into 20 long strips. On each strip glue 10 beans. If possible, involve students in making the bean strips. Have 40 loose beans available.

With the rest of the tagboard make two playing boards with columns labeled "Tens" and "Ones." Also prepare a card listing "Bonus" numbers. See Figure 2-13.

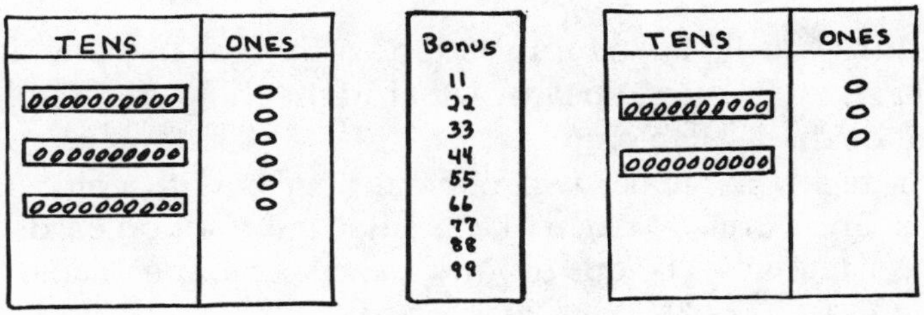

Figure 2-13

Directions:

1. Designate one or two students to be bankers. Banker has two functions: giving out beans and making exchanges.

2. Divide rest of group into two teams. (Game could also be played with two players, who would take turns acting as banker.)

3. A player on the first team throws the dice, then asks banker for the number of beans indicated by the sum of the dice. Beans are placed on playing board in appropriate columns.

4. Teams alternate, with different players on each team throwing the dice each turn and placing additional beans (equal to dice sum) on their board.

5. Whenever there are at least 10 beans in the "Ones" column, these beans should be exchanged for a bean strip, which would then be placed in the "Tens" column.

6. At the end of each turn, the teams should announce their new total. If, by chance, a team should have a total equal to any of the "Bonus" numbers, that team gets another turn. (It is possible for an entire game to be played without a "Bonus" total being reached.)

7. The first team to accumulate 100 or more beans is the winner.

99-BEAN SUBTRACTION

Summary: This game can be played before subtraction with exchanges has been introduced formally. Later, it can serve as a model. In this game two teams (or two students) pick cards as they attempt to get rid of all of their 99 beans.

Skills: Subtraction—model for subtraction with exchanging (also called borrowing, regrouping, or renaming)

Supplies: Tagboard, package of beans (at least 220), white glue, pen, paper. (Materials also may be used for "100 Bean Game.")

Preparation: Cut tagboard into 18 long strips (or use popsicle sticks). On each strip glue 10 beans. Let students help if possible. Have 20 loose beans available.

Using tagboard, make two playing boards with columns labeled "Tens" and "Ones." Next, make 12 small subtraction cards using any leftover tagboard, or cut up index cards or paper. Label the cards: 5,7,12,14,16,18,23,25,27,29,30,31.

Directions:

1. Divide group into two teams. Appoint one student to be banker. (This game also can be played with just two players.)

2. Each team sets up their playing board with nine bean strips and nine single beans for a total of 99 beans. (See Figure 2-14.)

3. The 12 subtraction cards are mixed and placed face down on the table.

4. A player on the first team picks a subtraction card and then proceeds to remove that number of beans from the team's playing board. If necessary, player will ask banker to exchange a bean strip for 10 single beans, before performing the subtraction.

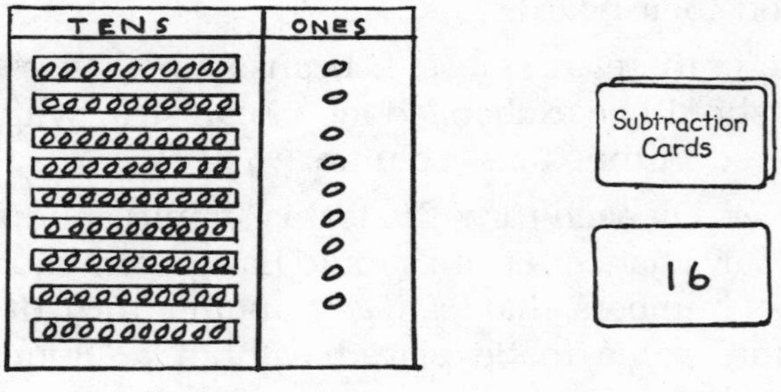

Figure 2-14

5. Teams alternate, with a different player from each team taking a card and removing the beans every turn.

6. First team to get rid of all 99 beans is the winner. It is not necessary to get the exact amount on the last turn to win; any larger number will do.

Note: At first, game should be played as is, with no paper or pencil. Later, after formal subtraction with exchanging (regrouping) has been taught, have students play again. This time do the subtraction on paper after each turn, thereby paralleling the actual exchanging and removal of the beans with the written subtraction example.

PAC-MATH

Summary: This lively board game for two players is a variation of the video game "Pac-Man." Students add two-digit numbers and these sums control the movement of Pac-Man and the four little monsters chasing him.

A variation is given for subtraction.

Skills: Addition (or subtraction) of two-digit numbers with regrouping

Supplies: Tagboard, pencil, scissors, and pens (several colors). Each player will need paper and pencil.

Preparation: To prepare gameboard, first draw a 15 by 15 grid with pencil. Then fill in circles and walls as shown in diagram (Figure 2-15).

Next cut out a circle-shaped Pac-Man and color him yellow, and four monster-shaped pieces to be colored with four different colors.

Finally, cut out 12 cards and label them with these numbers: 12, 15, 17, 24, 26, 28, 35, 37, 38, 43, 46, 49. (This game works best if these exact numbers are used on a 15 by 15 board.)

Directions:

1. The yellow Pac-Man piece is to be placed at the center of the gameboard, while the four monsters are each placed in a different corner of the board. The 12 cards are placed face down on the table.

2. One player will move Pac-Man, and the other player will move the four monsters who will be trying to catch Pac-Man.

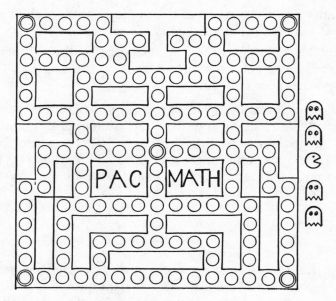

Figure 2-15

3. Each player draws a card and places it face up on the table. Using paper and pencil, both players add these two numbers. When finished, they check to see if they agree on the correct sum.

4. It is this sum that controls the number of spaces (circles) that the pieces may move: First, Pac-Man moves the number of spaces determined by the ten's digit of the sum. Next, the monsters move according to the unit's digit of this sum.

5. Only one monster may move on each turn, and the player controlling the monsters may choose any one of the four monsters to move.

6. Game continues and players draw another pair of cards which again are added. After sums are checked, Pac-Man always moves first, according to the ten's digit of the new sum, and then a monster will move, according to unit's digit. This time it may be the same monster as the one moved on the last turn, or it may be a new monster. (If the sum was 84, Pac-Man moves eight circles, followed by a monster who may move four circles.)

7. Pac-Man tries to avoid being caught by any monster. If, after all 12 cards are used (six additions) and Pac-Man has not been caught, then Pac-Man wins the game.

8. Monsters win if they manage to land on the circle that Pac-Man is on before the 12 cards are used up. The monsters need not win by an exact count. If, on the monster's turn, one of the monsters is four circles away from Pac-Man and the unit's digit is a 6, then the monsters win.

Note: Pac-Man moves from circle to circle horizontally, vertically or both, but not diagonally. Pac-Man is not allowed to go back onto a circle that he has just been on, in the same turn. However, on the next turn Pac-Man may go back over these spaces.

Variation: For subtraction, in place of the 12 addition cards, cut out 10 cards and place these subtraction problems on the cards:

72	91	50	82	81	76	85	63	98	80
−28	−19	−18	−29	−16	−24	−47	−37	−25	−39

Unlike addition, where two cards are to be used per turn, only one card per turn will be used for subtraction.

For each game, the 10 cards should be shuffled and six cards placed face down on the table. The other four cards are set aside. Before each turn a subtraction card is picked and both players should work out the subtraction. After the answers are checked, Pac-Man moves according to the ten's digit, and the monsters move according to the unit's digit. Again, Pac-Man wins if he is not caught by the monsters after six turns.

POST OFFICE

Summary: Letters, consisting of math problems, are to be delivered to the correct addresses, which contain the solution. Entire class checks for errors.

Skills: Addition or subtraction of two- and three-digit numbers

Supplies: Six to 10 envelopes, 12 to 30 index cards or slips of paper, pen, tape, paper and pencil for each student.

Preparation: Tape the envelopes to the walls around the classroom. On each is a different numerical address, such as 14 Elm Street or 935 Hillside Avenue. The streets should be real streets in the local area. Using index cards (or small sheets of paper) the teacher places one addition or subtraction problem on each card, making certain that the solutions correspond to the numbers in the addresses. For example, one card might read $63-49$; while another might be $282+496+157$. Since $63-49=14$, card goes to 14 Elm St; and since $282+496+157=935$, this card goes to 935 Hillside Ave.

Directions:

1. Choose students to deliver the mail. Give them a few cards each, which they must deliver by placing each in the proper envelope. Give them enough time to work out the problems.

2. Now the postal inspectors are selected, one for each envelope. They write the address on the chalkboard and under it, list each card found in that envelope. The lists are then discussed with the entire class and any errors are corrected.

Note: It's a good idea to label one envelope the "Dead Letter Office." Students place cards here when their answers do not correspond to any addresses. This could be because students have made mistakes, or because teacher has deliberately made up a few cards which have no corresponding addresses.

BEEHIVE

Summary: A game for two to six players. Players move through beehive and try to catch a certain number of bees.

Skills: Addition of three- and four-digit numbers, some subtraction

Supplies: Cardboard or tagboard, six index cards, pen, six different markers, paper and pencil for each player.

Preparation: Make a gameboard similar to the one in Figure 2-16, by first cutting a hexagon pattern out of cardboard. (See chapter on Geometry for instructions on making a hexagon.) Place pattern in center and trace hexagon; move pattern and trace again for each hexagon.
Cut index cards in half, making 12 small cards and label as follows: 2000, 2200, 2400, 2500, 2800, 2900, 3000, 3200, 3500, 3700, 3800 and 4000. Following each number write the word "BEES."

Directions:

1. Each player places a marker on a different "START" hexagon.

2. Small index cards are mixed up and one is given to each player. This represents the number of bees the player is to catch.

3. The object is for the player to achieve a score as close as possible to the sum on that card.

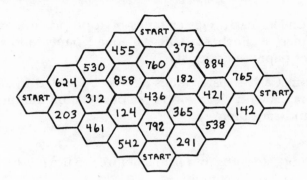

Figure 2-16

4. Each player makes one move per turn by moving marker a single space to any adjacent hexagon.

5. The first time marker lands on a particular hexagon, the player writes the three-digit number in that hexagon on scoring sheet. This represents the number of bees caught. On subsequent turns, the number is written down and added to the player's previous sum.

6. A player cannot land on the same number more than one time in a single game unless there is no other choice. However, other players may use this number. (Each individual's scoring sheet contains a list of numbers used by that player.)

7. When a player's total is equal to or greater than the number on the card, the player stops, removes marker from beehive, and circles total score.

8. Other players continue, taking turns, until each one of them finishes.

9. Now, all the players subtract the number on their individual cards from their total scores. The player whose total score is closest to the number on that player's card is the winner.

Note: If you wish to play again, all players should start at a different hexagon, thereby encountering a different series of additions in the next game.

SUPERMARKET

Summary: Totals are removed from cash register tapes and students working in teams add long lists of numbers to find correct totals.

Skills: Addition of four or more two- and three-digit numbers

Supplies: Collect cash register tapes of appropriate lengths from the supermarket. Each student will need paper and pencil. (Clear contact paper is optional.)

Preparation: Cash register tapes should be sorted by size. (You may cover tapes with clear contact paper.) Cut off the totals from each tape and save them.

Directions:

1. Arrange students into teams of three to five students. Each team should have a captain.

2. Give every student paper and pencil.

3. Give each team a few tapes of varying lengths.

4. Place all of the totals (which have been removed) on a separate desk in front of the room.

5. The captain of each team decides how to divide work. (For very long tapes, work may be broken down into subtotals.)

6. When a team has found a final total for one of their tapes, one of the students goes up to the front desk, announces the sum, and looks for the tape total that matches. If sum is correct there will be a match. If not, student must return to his or her team and try to find the error.

7. First team to match all their tapes with the correct totals wins.

3

Successful Permanent Games to Make for Multiplication and Division

These diverse multiplication and division games are designed to be fun. They have all the elements of popular commercial games—skill, chance, excitement, playing decisions and strategies. When used in conjunction with your present math curriculum they will develop and improve concepts, strengthen skills, enhance quick recall, and allow you to meet the individual needs and abilities of all your students.

Some of the games are to be played as whole class activities, while others are best played by smaller groups. To enable students to be self-sufficient when playing in groups, you should provide answer keys or multiplication tables whenever possible.

Games such as "Arrays," "Multi-Tac-Toe," solitaire "Hexagons," "Divide Away," and "Dungeons" are initially beneficial for students of all ability levels. However, as individual differences become apparent, these games are especially useful when continued by students with lower abilities. In addition, you might use the multiplication or division variation of "Top Card" or "Picture Puzzle" from the preceding chapter. As enrichment for your more able students, the following games combine math facts with skill and strategy: "Spin and Choose," "Reach the Top," "Multiplying Checkers," "Roman Chase" and "Division Grab." Games like "Journey to the Planets," "Once Around the Block," "Multiplication Touch" and "Dungeons" will allow you to evaluate your students' present abilities; while "Division Power," "Remainder Sale" and "Product Power" provide practice in more advanced work involving three- and four-digit numbers.

You might wish to supplement these multiplication and division games with puzzles, math tricks, card games, historic methods, and Quick-and-Easy games to be found in other sections of this book.

ARRAYS

Summary: This game can be played before formal multiplication has been introduced. It can serve as a model, and later be played to reinforce simple multiplication facts. Players choose which arrays they wish to make. The object is to use as few counters as possible. For two players or two small teams.

Skills: Multiplication using rectangular arrays

Supplies: Sheet of construction paper or tagboard, pen, scissors, eight index cards, and 44 counters (chips, buttons, or beans). *Note:* A checkerboard may be used in place of a 5 by 5 gameboard.

Preparation: Draw a 5 by 5 grid in the center of a sheet of construction paper as in the diagram below. Cut the index cards in half and label 15 card-halves. If appropriate use both a description and a pair of factors on each card. For example, for a 5 by 4 array, both "5 rows with 4 in each row" and "5 × 4" can be written. Make a card for each of the following: 2 × 3, 2 × 4, 2 × 5, 3 × 2, 3 × 3, 3 × 4, 3 × 5, 4 × 2, 4 × 3, 4 × 4, 4 × 5, 5 × 2, 5 × 3, 5 × 4, 5 × 5.

Directions:

1. Two players each receive five cards, which are then placed face up in front of the player. Set remaining cards aside.

2. Place four counters on the gameboard to form a 2 by 2 array as in Figure 3-1. Give 20 counters to each player. The object is to keep as many of these counters as possible.

3. Players alternate. First player looks at cards and chooses which array he or she would like to make and discards that card.

4. Players should always begin by using the counters already on the board, and then use their own counters if necessary to

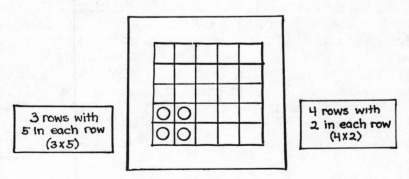

Figure 3-1

complete their chosen rectangular array. If this array calls for less counters than are on gameboard, player keeps the excess.

5. After the players each use all five of their cards, the one having the greatest number of remaining counters is the winner.

Example: There is a 3 by 5 array (15 counters) on the gameboard, player chooses a card describing a 4 by 3 rectangle. Player makes this array and keeps the three extra counters. Next player chooses a card asking for a 5 by 4 array. This player will use the 12 counters already on the board, and then will add eight of his or her own counters. If, on the next turn, a 4 by 5 array is to be made, no additional counters are needed. $(5 \times 4 = 4 \times 5)$

Note: Arrays can also be played with two teams of up to 5 players. Each time the team has a turn, a different player chooses which card to use and forms the appropriate array.

MULTI-TAC-TOE

Summary: This is a series of easy-to-make multiplication games where in each individual game, one factor remains constant throughout the game; for example, one game would involve only multiples of 3, while another would contain multiplication by 5. Each game is for two players, and is a variation of the familiar tic-tac-toe game.

Skills: Multiplication

Supplies: For each game—six index cards, 10 markers (five of one color and five of another color), pen.

Preparation: For each game—Leave one index card whole, to be used as a small gameboard, and cut the remaining five cards in half to be used as factor cards. Select the number to be used as a constant factor and prepare nine half-cards as follows. If, for example, 5 is to be used as the factor, label the nine cards: 1×5, 2×5, 3×5, 4×5, 5×5, 6×5, 7×5, 8×5, and 9×5. Prepare the small gameboard using the nine products corresponding to the factors on the cards. See Figure 3-2.

Of course, if a different constant factor is used, both the nine factor cards and the gameboard need to be changed.

Directions:

1. Place the nine factor cards face down in one pile.

2. Each player receives five markers of one color.

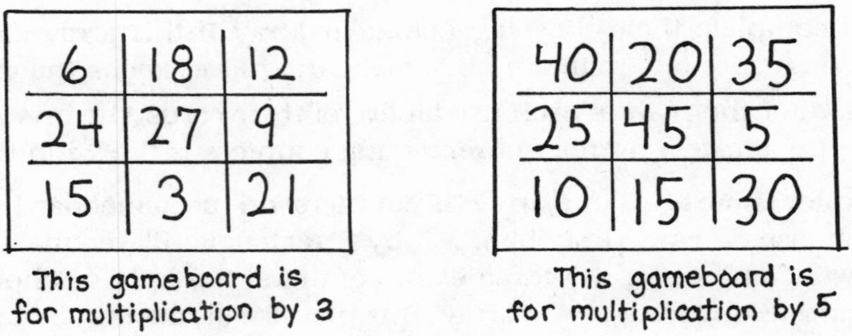

6	18	12
24	27	9
15	3	21

This gameboard is
for multiplication by 3

40	20	35
25	45	5
10	15	30

This gameboard is
for multiplication by 5

Figure 3-2

3. First player turns over top card and reads the two factors, such as "4 × 5," and places a marker on the product.

4. If the other player agrees that the correct product has been chosen, the marker remains. However, if the other player finds the product to be incorrect, the marker is removed and the factor card is returned face down to the bottom of the pile.

5. Game continues as players alternate turning over the top card and placing their markers on the correct products.

6. First player to get three in a row, horizontally, vertically, or diagonally wins.

Note: To allow the students to check each other's products, a separate card might be prepared for each game listing the nine corresponding multiplication facts.

JOURNEY TO THE PLANETS

Summary: A large chart of the solar system is placed in front of the classroom. Students select a spaceship, then go up to the chart and multiply the number on the spaceship with the number on each planet. Those who make successful trips, list their names on the back of the spaceships.

Skills: Multiplication

Supplies: Two large sheets of tagboard, 70cm × 56cm, (28in. × 22in.) of two different colors; felt-tip pens.

Preparation: Using one large sheet of tagboard, draw a diagram of the solar system similar to Figure 3-3, include both the planets' names and numbers.

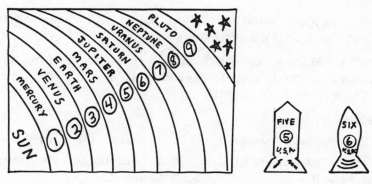

Figure 3-3

Cut the second sheet up into eight large shapes resembling spaceships, about 28cm × 10cm (11in. × 4in.). Number the front of each spaceship with a different number from 2 through 9. Draw lines across the back, one for each student in the class.

Directions:

1. Choose a student to come up to the chart and select a spaceship, noting the number on the front.

2. Using the spaceship as a pointer, student starts at the sun and points to each planet while multiplying the numbers on the planets by the number on the spaceship.

3. When students have correctly stated all nine products on their trip through the solar system, they are allowed to write their names on the back of the spaceship they used.

4. The object is for each student to eventually write his or her name on the back of each of the nine spaceships. This process may take several weeks, with the better students helping to prepare others for some of the more difficult spaceship trips (i.e., spaceships 7, 8, and 9).

REACH THE TOP

Summary: A game for up to four players. Students add, subtract, multiply and divide the two numbers on a pair of dice before deciding where to move.

Skills: Addition, subtraction, multiplication and division (optional)

Supplies: Tagboard or construction paper, two dice, self-adhesive labels, pens, four different markers.

Preparation: Make an enlarged copy of the gameboard in Figure 3–4. For each of the dice, write the numerals from 1 to 6 on the labels; cut to size and stick on the dice.

Directions:

1. Players place markers on "START" and first player rolls the dice.
2. The two numbers on the dice are read off and are then added, subtracted, multiplied and divided (if divisible).
3. If any of these results (+, −, × or ÷) are in a square adjacent to that player's marker, then player may move to that square. If not, player does not move that turn.
4. In this game an adjacent square is one which is to the right, left, up, down, or diagonal from the square student's marker is in.
5. Players alternate and first player to reach the top row is the winner.

Example: A player at start gets a 6 and a 2 from the dice. Since $6 + 2 = 8$, $6 − 2 = 4$, $6 × 2 = 12$ and $6 ÷ 2 = 3$, player can move to 3, which is in an adjacent square. However, on the next turn a 4 and a 3 are rolled on the dice. Player cannot move this time because 7, 1 and 12 are not in squares adjacent to the 3 square ($4 + 3 = 7$, $4 − 3 = 1$, $4 × 3 = 12$). On the third turn a 6 and a 4 turn up and player chooses to go from the 3 to the 2 square.

REACH THE TOP

✳ 5	✳ 12	✳ 6	✳ 9	✳ 4
8	11	1	18	10
2	4	8	3	12
10	30	5	2	7
25	3	4	1	20
15	6	2	0	9
7	1	5	3	6
16	24	START	10	36

Figure 3-4

MULTIPLYING CHECKERS

Summary: A checkers game with numbers on each checker. When checkers are jumped, student receives a score equal to the product of the two numbers. Winning strategies will have to take into account how to set up the checkerboard and which numbers to move first.

Skills: Multiplication

Supplies: Set of checkers and checkerboard, red and black pens, self-adhesive labels or paper and tape.

Preparation: On 24 small tabs of paper write 12 numbers, once in black ink and again in red ink. You can use the numbers 1 through 12, or if desired, use only a few numbers repeated many times as in Figure 3-5. Tape these to the top of the checkers.

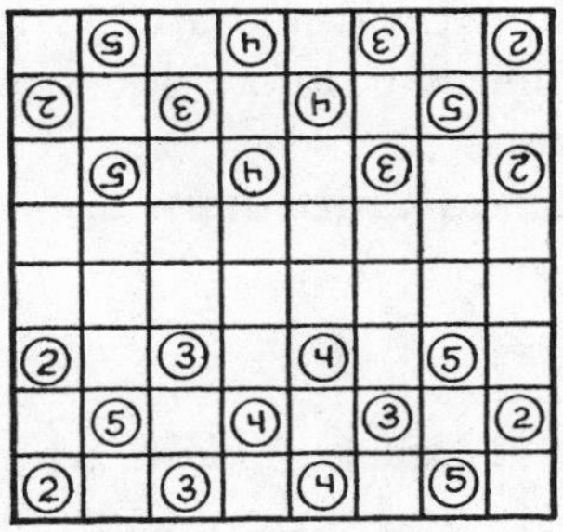

Figure 3-5

Directions:

1. Set up checkerboard as usual, placing numbered checkers in any order desired.

2. Players use the rules of checkers with this one difference: each time one player's checker jumps another, the numbers on top of these two checkers are multiplied and the result is the score of this player. The jumped checker is then removed.

3. For example, if red 3 jumped black 5, red would receive 15 points.

4. For each new jump, the two numbers are multiplied and the result added to player's previous score.

5. The player with the highest score at the end of game is the winner; or, players may select a goal, such as 200 points, and the first one to reach this total wins.

Optional: When a checker becomes a king, its value doubles. For example, a "4" checker that becomes a king now has a value of "8".

HEXAGONS

Summary: A challenging game, excellent for reviewing multiplication facts. Can be played by one student as solitaire, or by a small group as a competitive game.

Skills: Multiplication

Supplies: Tagboard and felt-tip pen.

Preparation: First prepare a hexagon using a small piece of tagboard. Make the hexagon at least 10 cm (4 in.) in diameter. See chapter on Geometry for instructions on making a hexagon. Trace 12 hexagons, arranging them as in Figure 3-6. Label each side of the hexagons so that adjacent sides represent corresponding factors and products. Products are underlined.

Note that the products on extreme right match factors on the extreme left, and top products and factors match bottom factors and products. This will allow for numerous variations when rearranging hexagons.

Cut out all 12 hexagons, mix them up and stack in a pile.

Directions:

For Solitaire:

1. Player takes the 12 hexagons and arranges them as a puzzle so that all the factors and products on adjacent sides match.

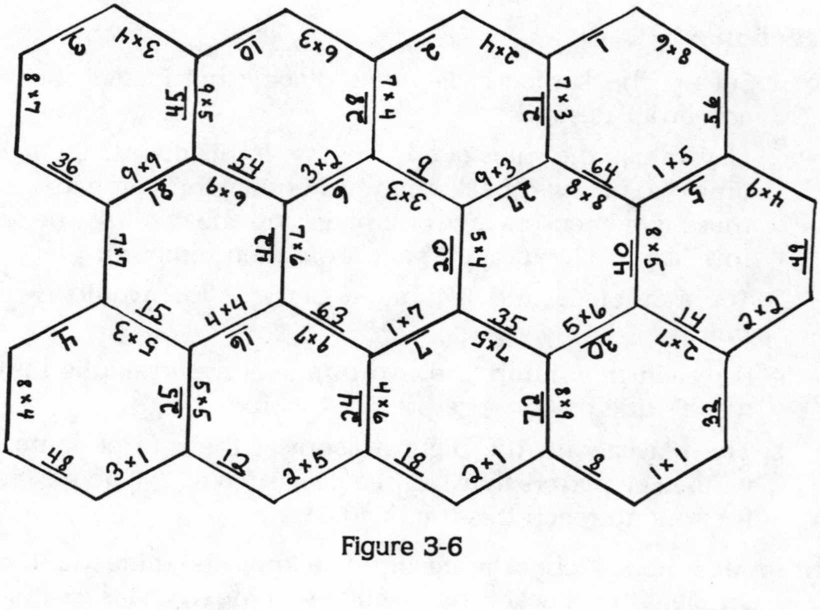

Figure 3-6

2. The solution does not have to be identical to Figure 3-6; there are numerous arrangements which are also correct.

For two to four players:

1. Before playing, each player takes one hexagon and the player with the hexagon containing the highest product goes first.

2. To begin the game, all hexagons are given out to the players: six each for two players, four each for three players, or three each if four play. These hexagons should be placed face up in front of each player.

3. The first player places one hexagon in the center of the playing area and receives one point for going first.

4. Players now take turns placing one hexagon on each turn in the center area so that adjacent sides contain corresponding factors and products.

5. Each time a player correctly places a hexagon so that adjacent sides match, player calls out this math fact, such as "7 × 6 = 42," and receives one point for each correct match. If no such match can be made, player passes.

6. As the game progresses it will be possible to make more than one correct match in a single turn. A player may place a hexagon so that two or three or even four sides match correctly with sides of adjacent hexagons. Player would then call out all of these math facts and receive one point for each.

7. When all hexagons have been placed in the center, player with the most points is the winner.

ONCE AROUND THE BLOCK

Summary: A multiplication game for two to four players. Students move from house to house as they attempt to be the first to go around the block. Students multiply numbers and move to adjacent houses only when the product of these numbers can be found inside that house.

Skills: Multiplication (using 4 through 9 as factors)

Supplies: Tagboard or construction paper, two dice, self-adhesive labels, pen, four different markers.

Preparation: Prepare a large gameboard with a diagram similar to the one in Figure 3-7. For each of the dice, use self-adhesive labels and write the numerals 4 to 9 on the labels. Then cut to size and stick on both dice.

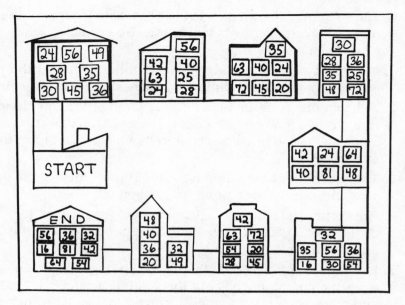

Figure 3-7

Directions:

1. All players place markers on the house marked "START."
2. First player throws dice and finds the product of the two numbers. If this product is in the next house, player moves his marker to this house; otherwise, player does not move.
3. Players take turns each time throwing the dice and moving only if product is found in the next adjacent house.
4. Winner is the student who moves around the block, going through each of the nine houses in the correct order.

Note: A multiplication table should be made available to enable students to check each other's math facts.

SPIN AND CHOOSE

Summary: A chalkboard game involving the entire class. This game centers around the use of three numbers, selected in advance, used as multipliers. Each game provides practice with 30 multiplication facts.

Skills: Multiplication

Supplies: Chalkboard, chalk, spinner (paper plate or circular piece of tagboard, paper clip and brass fastener).

Preparation: In advance, prepare spinner as in Figure 2-11. Brass
fastener and opened paper clip should be loose enough for paper
clip to spin freely. Label the 10 sections from 1 to 10.

Immediately before the game, draw a 5×6 grid on the chalkboard.
Select three numbers to be used as multipliers in this game. A
chart listing all multiples of these three numbers should be placed
on the chalkboard. The following example will use the numbers 4,
5, and 6; however, other numbers could just as easily be used.

4	5	6
8	10	12
12	15	18
16	20	24
20	25	30
24	30	36
28	35	42
32	40	48
36	45	54
40	50	60

This would be a good time to review these multiplication facts with the
students. Then these 30 numbers should be placed in a random order
inside the 30 squares of the 5×6 grid. Figure 3-8 shows the grid with
multiples of 4, 5 and 6.

Directions:

1. Divide class into two teams. Each team will choose a letter to be
   used for identification, such as A or B.

2. A player from the first team comes up to the chalkboard, spins
   the spinner, and calls out the number the spinner has landed
   on.

3. Now the team will choose one of the 30 squares to mark. There

60	28	12	40	45	12
30	6	24	10	30	25
48	8	50	24	16	42
20	36	54	15	35	4
18	40	5	20	36	32

Figure 3-8

will be three choices. This spinner number may be multiplied by any of the three multipliers chosen at the beginning of the game. If the spinner had landed on an 8, then in the sample game the team could choose $8 \times 4 = 32$ or $8 \times 5 = 40$ or $8 \times 6 = 48$ and place their team's letter in the square with the chosen product. (These three math facts might be placed on the chalkboard if the teacher felt it would be useful.)

4. Choose another student from this team to make the decision and mark the selected square on the board with the team letter. Only one square may be selected on each turn.

5. Teams alternate with at least two players from each team participating on every turn. If it should happen that on a particular turn all three product squares have already been chosen; then that team will not mark any square on that turn.

6. The first team to get four in a row wins—horizontally, vertically, or diagonally. If this proves too difficult, teams might try to get only three in a row.

MULTIPLICATION TOUCH

Summary: A multiplication table approach to learn multiplication facts through 9 × 9. A game for four players.

Skills: Multiplication

Supplies: Large piece of tagboard, pen.

Preparation: On the large tagboard draw a square 27cm × 27cm. Divide square into nine rows and nine columns and label with numerals 1 to 9, as shown in Figure 3-9. Leaving a small border around the large square, cut out the gameboard.

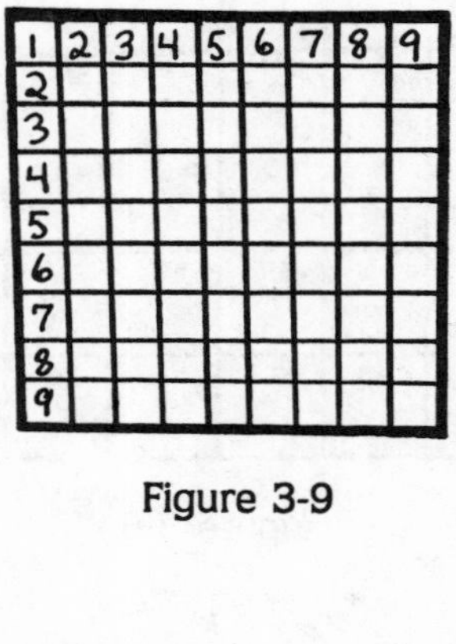

Figure 3-9

With remaining tagboard, cut out 64 square tiles, each slightly smaller than the 3cm × 3cm squares on the gameboard. Consider the gameboard as a multiplication table and the small tiles as the products which belong in the 64 blank squares. Label these 64 tiles with the appropriate products from 4 (=2×2) to 81 (=9×9). Draw a small multiplication table modeled after the gameboard. It will be useful both in the preparation and the playing of the game.

Directions:

1. The 64 tiles are placed face down and each player selects seven tiles.

2. After the first player places a tile on the gameboard in the proper square, other players must use only tiles that can be placed in squares touching tiles already on the board. (Tiles with corners meeting are considered touching.)

3. For example, suppose the first player should place the tile with product "15" on the square where the row labeled 3 meets column labeled 5. The next player would need to have one of the seven adjacent product tiles to be allowed to place a tile on the board: 8, 10, 12, 16, 18, 20, 24.

4. On each turn, a player may place only one tile on the board. If the player cannot place a tile so that it touches a tile on the board, or if player places a tile incorrectly, he or she must take an extra tile from the group of face-down tiles on the table.

5. The first player who uses up all his or her tiles wins the game.

PRODUCT POWER

Summary: A thought-provoking game in which students eagerly multiply two three-digit numbers and examine the results. This game can be played by two players, or modified for two teams to involve the entire class.

Skills: Multiplication of two three-digit numbers

Supplies: Deck of cards or nine blank cards cut from index cards or tagboard, paper and pencils.

Preparation: If a deck of cards is available, use only nine cards—the ace through the 9 card in any single suit. If not, cut out nine blank cards and label them 1 through 9.

Directions for two players; modify for team play.

1. Take the nine cards and place them face down. Both players take three cards and turn them over.
2. Each player arranges those three cards into a three-digit number.
3. The three-digit number of the first player is to be multiplied by the three-digit number of the second player.
4. Both players should work out the multiplication with paper and pencil; then they should compare and check work for accuracy.
5. Scoring is done as follows: players receive a point each time any of their three original numbers appear in the partial products or in the final product. (See sample game.)
6. After a few rounds, player with the most points wins.

Sample Game: Suppose first player picks cards with 1,3, and 5, and arranges them to form the three-digit number 513. Meanwhile, the second player chooses cards 2, 4, and 7, and arranges them to form 724. Both players now multiply 513 by 724.

```
       FIRST PLAYER           SECOND PLAYER
              5 1 3                    5 1 3
              7 2 4                    7 2 4
        2 0 5 2-partial product ------ 2 0 5 2
      1 0 2 6 ---partial product --- 1 0 2 6
      3 5 9 1 ---partial product --- 3 5 9 1
      3 7 1 4 1 2--final product -- 3 7 1 4 1 2

     SCORE = 8 points        SCORE = 6 points
```

Note: If players had chosen different orders, results would have been different.

```
              1 3 5        First player would
              7 4 2        receive 4 points.
              2 7 0
            5 4 0          Second player would
          9 4 5            receive 5 points
        1 0 0 1 7 0
```

Who would receive the higher score if 531 was to be multiplied by 427?

DIVIDE AWAY

Summary: This game, using counters and an empty egg carton, serves as a model for division. It is not necessary, however, to introduce formal division before playing. For two players or two teams.

Skills: Division

Supplies: Egg carton, dice (one die), 25 small counters, paper and pencil.

Preparation: Take an empty egg carton and remove cover. Using the bottom half, place numbers from 1 to 12 in the 12 sections, as in Figure 3-10(a).

Directions:

1. Place the 25 counters in the center of the playing area.
2. First player throws a single die and calls out the numeral on top. This numeral represents the number of counters that are to go into each section of the egg carton. If this number is 1, however, player should throw die again.
3. Suppose the number on the die is 3; then three counters are placed into the first section, three more into the second section, three in the third, and so forth until there are less than three counters left. See Figure 3-10(b). Eight sections have been filled and one counter remains. $(25 \div 3 = 8r1)$
4. The player's score would be the number of sections properly filled (in the example, this is eight), and the remaining counter is kept by the player to be turned in at the conclusion of the game.
5. The game continues with the 24 counters in the egg carton placed in a pile in the center of the playing area. The second player throws a 1 on the die. Since a 1 turned up, this player throws again. Suppose a 5 turns up. Then the player would place

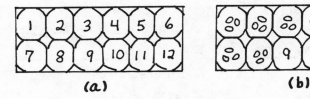

(a)   (b)

Figure 3-10

five counters in each section of the egg carton until there are less than five counters left. Four sections would be filled with five counters each, and four counters would remain. ($24 \div 5 = 4r4$) The second player would receive a score of four points and keep the remaining counters.

6. There are now 20 counters left in the game, which are again placed in a pile in the playing area. Play continues with players receiving a score equivalent to number of sections filled and keeping any remaining counters. At the conclusion of the game, players trade in remainder counters for points. Each group of four remainder counters equals one point.

7. The game ends whenever the number on the die is larger than the number of counters left in the game. These few counters are left in the playing area and are not taken by the players.

8. Now players turn their attention to the score. After the remainder counters have been traded in, the score is totaled and the highest score wins. A complete sample game between two teams is provided below. Although division notation is used here, it is not necessary to use this notation when first introducing the game.

Sample Game:

TEAM A	TEAM B
$25 \div 2 = 12$ r1 ($25 - 1 = 24$)	$24 \div 3 = 8$
$24 \div 6 = 4$	$24 \div 5 = 4$ r4 ($24 - 4 = 20$)
$20 \div 6 = 3$ r2 ($20 - 2 = 18$)	$18 \div 3 = 6$
$18 \div 2 = 9$	$18 \div 4 = 4$ r2 ($18 - 2 = 16$)
$16 \div 5 = 3$ r1 ($16 - 1 = 15$)	$15 \div 6 = 2$ r3 ($15 - 3 = 12$)
$12 \div 2 = 6$	$12 \div 4 = 3$
$12 \div 3 = 4$	$12 \div 5 = 2$ r2 ($12 - 2 = 10$)
$10 \div 6 = 1$ r4 ($10 - 4 = 6$)	$6 \div 3 = 2$
$6 \div 2 = 3$	$6 \div 4 = 1$ r2 ($6 - 2 = 4$)
$4 \div 5$ cannot divide	

points 45 r8   points 33 r13

Point Score = 45   Point Score = 33

Remainder Score = $8 \div 4 = 2$ Remainder Score = $13 \div 4 = 3$ (r1)

Total Score = $45 + 2 = 47$ Total Score = $33 + 3 = 36$

DUNGEONS

Summary: A separate board should be made for each number from 3 to 9. Figure 3-11 shows a board for division by 7. Up to four students go through the dungeon by dividing the number in the squares by the given number. (In this example, 7.)

Gameboard may also be used for a multiplication game. See variation.

Skills: Division (or multiplication)

Supplies: Tagboard, pens, four different markers, one die (dice).

Preparation: Prepare a separate gameboard for each number desired. Below is a board to be used for division by 7. (For a board for division by another divisor, say 8, change layout of gameboard and change multiples of 7 to multiples of 8.)
For each game, provide an Answer Key.

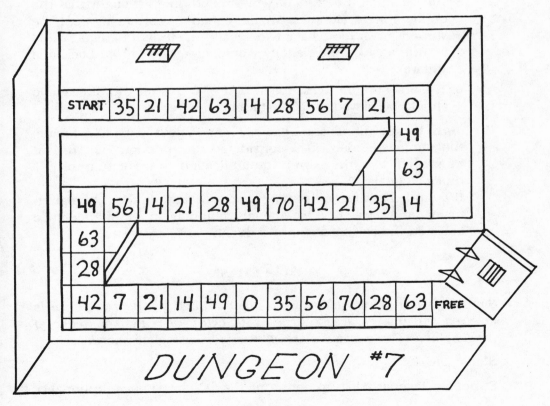

Figure 3-11

Directions:

1. Students place markers on "START." Each student, in turn, throws the die and moves marker forward, according to the number on the die.

2. When student's marker lands in a square, the number in that box must be correctly divided by the game number. In this case, 7.

3. Suppose marker lands in square 42. Student must state, "42 divided by 7 is 6." If the division is stated correctly, the student remains in that square.

4. If incorrect, student is told the correct quotient (from Answer Key) and as a penalty he or she must move back a few spaces. The exact number of spaces for the penalty is the difference between the student's incorrect answer and the correct quotient.

5. For example, if a student who landed on 42 thought that the quotient was 8, since 8 is two more than the correct quotient, the student must move back two squares. If student had guessed 5 for the answer, the penalty would have been to go back one square.

6. The first student to escape to "FREEDOM" wins. However, game should be played until all students are free.

Variation: For a multiplication game, when student lands in a square, the number landed on is considered the product. The student must then give the correct multiplication fact. For example, "7 times 6 equals 42."

If a student makes an error, the correct answer should be given (using an Answer Key). The penalty could again be the difference between the correct factor and the student's answer.

ROMAN CHASE

Summary: Two players chase each other around the board; choice is involved and players are rewarded with extra turns for knowledge of division facts. Three divisors are selected for each gameboard.

Skills: Division

Supplies: Tagboard, felt-tip pens, one die. (12 checkers are optional.)

Preparation: To make a game involving division by 4, 5, and 6 use numbers given. If other divisors are desired, see note at the end of this game.

Cut rectangle at least 36cm × 24cm, out of tagboard. If checkers are available use six red and six black, then place numbers on small pieces of paper and paste on the checkers. In place of checkers, 12 small, circular discs cut from tagboard may be used. They should be colored two opposing colors, and a number written on each. Label the 12 discs with 4, 5, and 6 (each four times), as shown in Figure 3-12.

Gameboard contains two rows of 12 long rectangles, each with a number along the edge representing multiples of the three numbers on the discs. The 24 numbers can be copied from Figure 3-12.

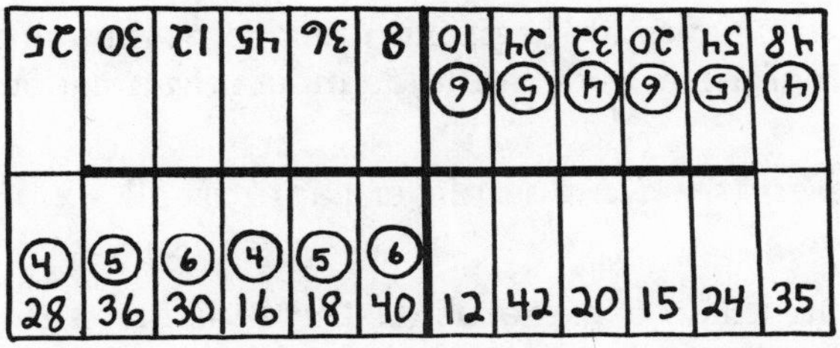

Figure 3-12

Directions:

1. Set up gameboard as shown. The two players sit opposite each other, each having six discs of the same color together on his or her side of the board.

2. Players, in turn, move counterclockwise around the board, according to the throw of a single die. The aim of the game is to land the disc in a space occupied by the opposite player, thereby capturing it.

3. Players are rewarded for their knowledge of division. After the initial throw of the die, players may choose any of their own discs to move; and if they choose a disc that lands in a space whose number is perfectly divisible by the number on that disc, then the player may take another turn. But now the player no longer has a choice of disc and must move the same disc again. This may be repeated until disc lands on a space into which it is not perfectly divisible.

4. To illustrate the above rules, suppose first player throws a 4 on the die and may move any disc four spaces to the right.

   If player had moved the disc with divisor 6 on it on space 40 (see Figure 3-12) four spaces forward, it would have landed on a 15, which is not divisible by 6. Instead, the player should choose to move the adjacent 5 disc (which is on space marked 18) forward four spaces so that it lands on a 20. Player sees that 20 is divisible by 5 and announces "20 ÷ 5 = 4" and then takes another turn.

   After throwing a 6 on the die, player continues to move this disc six spaces counterclockwise: 20-15-24-35-48-54-20. Not only is this space divisible by 5 (the disc number), but it contains an opponent's disc, which is captured and taken off the board. Player may go once again until disc lands on a space not divisible by 5.

5. Each time players begin another turn, they are free to choose which of their discs would be most advantageous to use for that turn.

6. Game ends when one player captures all six of the opponent's discs.

Note: You can make a game using three different divisors by taking whatever numbers you desire and placing them on the 12 discs. Next, find the products formed by multiplying these three divisors by the numbers 2 through 9. 24 products should be formed, which should be placed randomly around the gameboard.

DIVISION GRAB

Summary: An interesting game in which students choose their own divisors, often gaining insights into divisibility. For two to four players.

Skills: Division of a three-digit number by a one-digit divisor

Supplies: 32 index cards, felt-tip pen, and paper and pencil for each player.

Preparation: Cut eight index cards in half. On the unlined side of the 16 card-halves, write the numerals 2 through 9, using each numeral twice.

On the whole index cards, write the following numbers, one per card: 144, 147, 189, 217, 245, 270, 296, 336, 343, 375, 387, 441, 460, 475, 486, 504, 512, 522, 567, 648, 720, 729, 873, 920.

Directions:

1. Select one player as the dealer, and another as scorekeeper. A game consists of at least four rounds, after which the score is totaled.

2. The 16 small cards containing the divisors are to be spread out face up in the center of playing area.

3. The 24 larger cards are to be placed in one face-down pile and given to the dealer. The dealer then places one large card face down in front of each player.

4. On the dealer's signal of "Over," all players turn over their cards and look at their own three-digit number. Each player now grabs for a divisor card; players should try to choose the divisor they believe will give them the best score, according to the following rules.

5. To determine the number of points each player receives: first the players, using paper and pencil, each divide their three-digit number by the divisor number chosen. If their number is perfectly divisible by this divisor, leaving no remainder, then they receive 10 points. However, if it is not perfectly divisible and there is a remainder, then the player will receive a score equal to that remainder. Examples: $147 \div 3 = 49$, score = 10 points. $296 \div 6 = 49$ r2, score, 2 points.

6. All cards used in the first round, both large and small, are collected and removed from the game. Dealer now gives each player a new face-down card. Game continues and players again grab for new divisor cards, but on each round there are fewer of these cards to choose from.

7. Players may check on and challenge the other players' division; if an error is found, then that player would receive no points for that round.

8. After four or five rounds (four rounds if four play, five rounds if two or three play), the score is totaled and the highest score wins.

Note: All of the numbers given are divisible by at least one of the divisors; and many are perfectly divisible by three or four different divisors.

Remind students that a remainder cannot be larger than the divisor. Of course, the best score (10 points) will be received by those who have numbers perfectly divisible by chosen divisor. However, if a student is not able to choose such a divisor, it might be to the student's advantage to choose a larger divisor (say 9 rather than 3) which might leave a higher remainder. Allow students to develop their own strategies.

DIVISION POWER

Summary: For two teams. Players arrange five digits into division problems, trying different arrangements of these digits as they compete for the highest score.

Skills: Division of three-digit numbers by two-digit divisors

Supplies: 10 blank cards cut from index cards or tagboard, pen, paper and pencil.

Preparation: Label the 10 cards with the numbers 0 to 9.

1. Mix cards and give five cards to each team.
2. Each team now has five numbers. A division problem is to be made from these five digits, using any two for the divisor and the other three for the dividend. The order of the digits is up to the players.
3. After the division is worked out, look at each digit to determine score. One point is given each time one of the original digits appears. (See following example.)
4. Give teams time to try out a few arrangements before committing themselves to a particular order.
5. Teams should check each other's work. Highest score wins.

Example:

Team A—digits 1,2,4,6,7   Team B—digits 0,3,5,8,9

```
        29                          28
   24 / 7 1 6                  35 / 9 8 0
      4 8                         7 0
      2 3 6                       2 8 0
      2 1 6                       2 8 0
          2 0 Remainder                0 Remainder
```

Team A—13 points   Team B—12 points

REMAINDER SALE

Summary: Up to four teams solve difficult division problems having remainders. These remainders allow teams to buy items on sale.

Skills: Division of four-digit numbers by two-digit divisors

Supplies: Pen, 16 pictures cut out of magazines, paste, 48 index cards (if possible, 16 of these should be a different color).

Figure 3-13

Preparation:

1. Cut out pictures that will be of interest to students and paste them on the 16 index cards. Sixteen prices, which correspond to the remainders of the division examples, should be written on the cards: $5, 7, 8, 9, 10, 12, 13, 15, 18, 24, 25, 28, 32, 36, 41, 46.

2. On the 32 plain index cards, write the following division examples, one per card. Do *not* include the answers on these cards.

$15\overline{)1419} = 94r9$	$35\overline{)1459} = 41r24$	$50\overline{)1857} = 37r7$
$24\overline{)1165} = 48r13$	$36\overline{)1717} = 47r25$	$52\overline{)4362} = 83r46$
$26\overline{)1209} = 46r13$	$37\overline{)1578} = 42r24$	$54\overline{)1573} = 29r7$
$28\overline{)3536} = 126r8$	$37\overline{)8579} = 231r32$	$54\overline{)3466} = 64r10$
$31\overline{)3945} = 127r8$	$38\overline{)8783} = 231r5$	$67\overline{)3798} = 56r46$
$31\overline{)3996} = 128r28$	$42\overline{)8978} = 213r32$	$72\overline{)3924} = 54r36$
$31\overline{)9852} = 317r25$	$42\overline{)9165} = 218r9$	$73\overline{)2646} = 36r18$
$32\overline{)943} = 29r15$	$42\overline{)10127} = 241r5$	$75\overline{)2562} = 34r12$
$32\overline{)4348} = 135r28$	$45\overline{)6161} = 136r41$	$82\overline{)1734} = 21r12$
$34\overline{)7626} = 224r10$	$47\overline{)3026} = 64r18$	$82\overline{)3726} = 45r36$
$35\overline{)995} = 28r15$	$47\overline{)6715} = 142r41$	

3. Prepare an extra card or piece of paper containing all 32 division problems, including the answers. Note that the above list of problems is arranged so that the divisors are in numerical order. This "Answer Key" will be used for checking answers later in the game.

Directions:

1. Choose one student to be the leader and divide the rest of class into two to four teams.

2. Spread out the 16 picture cards of items for sale face up on a table so that all are showing.

3. The leader sits at the table with the 16 picture cards, the Answer Key, and a pile of 32 problem cards. He or she gives out one problem card to each team.

4. Each team works together copying over their own problem and solving it. When they are done, one student from the team goes up to the leader who checks the correct quotient and remainder from the Answer Key.

5. If answer is correct, student looks over the table for an item that is on sale and whose price matches the remainder. If the correct answer was 45r36, then the team could buy the boat whose sale price was $36.

6. Player takes the $36 picture card, and keeps it by the team. Later in the game, it may be possible that the card equal to team's remainder is not available. This is possible because there are two problems that have the same remainder (observe previous list) and another team may have taken this picture card already.

7. In either case, the team receives a new problem card from the leader, and goes to work to solve this problem. This time, another student from the team can go up to the leader to see if the proper picture card is available.

8. If the leader discovers that the work is incorrect, the team does not take a picture card, but instead has a choice. The team can decide to try to correct the problem or choose to get a new problem card.

9. Teams work independently of each other, as quickly and carefully as possible. The first team to collect five picture cards wins the game.

4

How to Make Fractions as Easy as Pie

All of our students have had concrete experiences with integers prior to the introduction of numerals and their basic operations. Fractional numbers are an even more difficult concept to learn; partly because students have had less experience with fractions, and partly because of the confusion often arising from the various interpretations of fractional numbers. In addition, there are many equivalent forms of each fraction, for example, ½, ¾, ⁵⁄₁₀, .5, .50, are all numerals which represent the same fractional number.

Certainly, the concept of fractions should be introduced through a variety of concrete experiences. Moreover, when students leave concrete models and begin to work with the abstract symbols for fractions, sometimes the meaning of the fractions involved is lost. For example, take a student attempting to add the symbol "¼" with "⅓," who relies not on any mental images of these fractions, but only on a confused series of memorized rules. Such a student may come up with as improbable an answer as "¼ + ⅓ = 12!" When students make such a mistake, it might be time to return to a physical model, such as "Fraction Bars."

Lower-grade students just being introduced to fractions might enjoy "Fraction Pie," "Move Along," and "Fraction Match." As students become ready to move from the concrete and semi-concrete into abstract fractional symbols, they might try "Make-a-Whole." Later, after learning operations with fractions, students will be ready for "Fraction War," "The Fraction Maze," and perhaps "Catch the Thief." An interesting game which helps students to develop insights into the meaning of fractions is "3-2-1 Pass."

"Decimal Pac-Math" is a special variation of the video game Pac-Man, where students add decimal fractions to control the moves of the pieces. Any problems involving decimals can be incorporated into "Decimal Racetrack," a game where the gameboard itself serves as a model for decimals; while "Equivalence Lotto" includes percentages as well as fractions and decimals. This chapter concludes with "The Jungle Game"—a board game more complex than checkers and almost as difficult as chess, where the score is originally a fraction and later converted to a decimal. While this game is not for all students, those with higher ability will find this game more challenging with repetition as individual players begin to develop their own strategies.

FRACTION PIE

Summary: Using circular models of fractions, students learn to recognize fraction parts as they put fractions together to form a whole. This game is for two players or may be adapted for team play.

Skills: Fractional numbers—fractions as parts of a whole

Supplies: Eleven 6-inch (15cm) paper plates or 11 circles cut from tagboard, one paper clip, one brass fastener, red and blue pens.

Figure 4-1

Preparation: Leave five paper plates whole. Label two with red pen (cherry pie plate) and two with blue pen (blueberry pie plate). Make the next plate into a spinner divided into eight parts and labeled as in Figure 4-2.

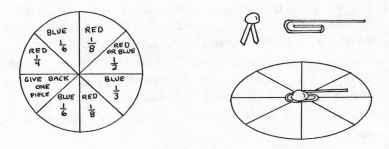

Figure 4-2

Then, attach the brass fastener through an opened paper clip and then through the center of plate—loosely enough to allow the paper clip to spin freely.

Cut six paper plates into fractions. Cut the first into two ½s (halves), the second into four ¼s (quarters), and the third into eight ⅛s (eighths). These pieces should be labeled in red ink with the

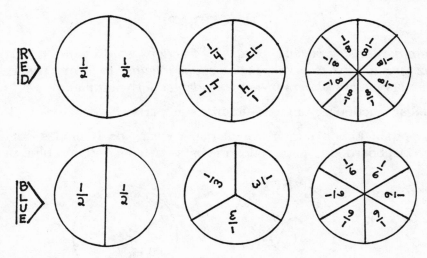

Figure 4-3

appropriate fractions (Figure 4–3). These are pieces of the red cherry pie.

The fourth plate should be cut into another two ½s (halves), the fifth into three ⅓s (thirds), and the last into six ⅙s (sixths). These will be the pieces of the blueberry pie and should be labeled in blue with the correct fractions.

Directions:

1. Both players are given one red and one blue pie plate.

2. Player spins the spinner and takes the fraction indicated on spinner (if needed and if available). It is placed in the proper plate: red fractions in red plate, blue fractions in blue plate.

3. Players alternate and the winner is the first to complete both whole pies—red and blue. Combinations of different fractions may be used to complete the pies.

4. The last piece must fit exactly to complete a pie. A few examples are below.

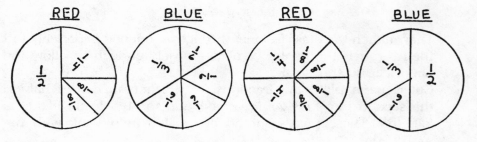

Figure 4-4

Variations: Allow for exchanging of equivalent fractions when necessary: ½ = two ¼s, ½ = three ⅙s, ½ = four ⅛s, ¼ = two ⅛s, and ⅓
   = two ⅙s. On spinner replace the second "⅙" and "⅛" with "lose
   ⅙" and "lose ⅛."

MOVE ALONG

*Summary: Player spins a spinner and moves along on the gameboard
until diagram of that fraction is reached. A game for up to four players.*

Skills: Recognizing fractions

Supplies: Felt-tip pens, tagboard, paper clip, brass fastener, paper plate
   or circular piece of tagboard, four different markers.

Preparation: Prepare a gameboard similar to that shown in Figure 4-5,
   showing fractions as parts of a whole or as parts of a set.
   Next, prepare a spinner by dividing a circle into eight parts and
   labeling them as shown. Next, attach brass fastener through an
   opened paper clip and then through center of circle—loosely
   enough to allow paper clip to spin freely.

Directions:
   1. All markers are placed on "START."
   2. In turn, players spin the spinner and move markers according to
      directions on spinner.

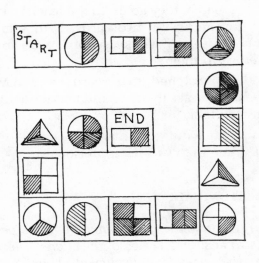
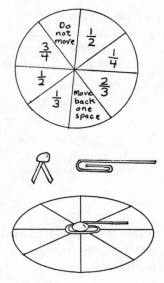

Figure 4-5

3. If spinner points to a fractional number, then player moves along the gameboard until a diagram of that fraction is reached.
4. First player to reach "THE END" wins.

FRACTION MATCH

Summary: For the entire class. Half the students have cards containing fractional numbers, the other half have pictures illustrating these fractions. Students race against time to match the fractions.

Skills: Recognizing fractions

Supplies: Index cards (one for each student), felt-tip pen, watch or clock for keeping time.

Preparation: The number of index cards should equal the number of students in the class. On half of these cards, write a fractional number; and on the remaining cards, draw pictures illustrating each of the fractions.

Directions:

1. The index cards should be distributed to the class, with each student receiving either a fractional number or a picture of a fraction.
2. Students move about the room looking for a match—a fraction plus the appropriate picture.
3. When students find such a match, they go up to the teacher (or an appointed leader) who checks the two cards. If they are correctly matched, teacher or leader will keep both cards and the two students will return to their seats.
4. After all the students have matched their cards and have returned to their seats, teacher notes the time and tells students how long the fraction match took.
5. Cards are now mixed up, given out, and matched again—the object now is to use less time then was used previously.

FRACTION BARS

Summary: Class is divided into two to six small groups. Students use rectangular models of fractions to answer a series of questions. These materials are also very effective when used by individual students.

Skills: Understanding fractions

Supplies: Construction paper* 12 in.× 9 in. (Six different colors), pen, paper, several pairs of scissors. (*Use six sheets for up to three teams; 12 sheets for six teams.)

Preparation: For each set of fraction bars cut six long rectangles 12 in. × 3 in., each a different color. Students might help in the preparation. Section fraction bars as shown in Figure 4-6 below, either labeled or left blank according to the abilities of the students.

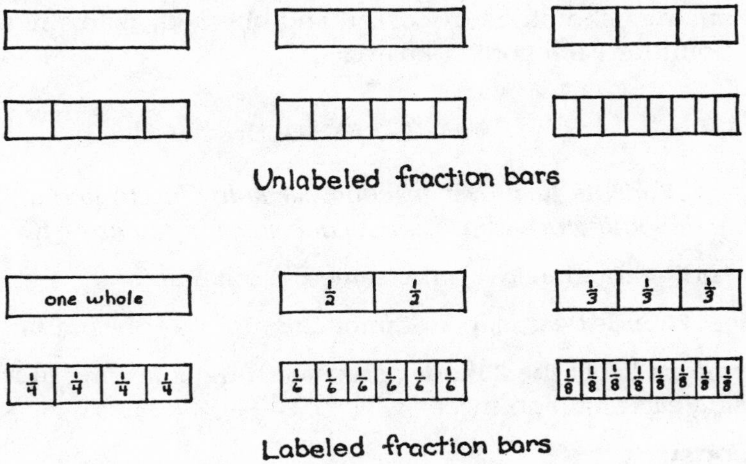

Unlabeled fraction bars

Labeled fraction bars

Figure 4-6

A list of 12 to 24 questions involving fractions should be prepared, and duplicated for each group. Following are some examples:

1) Which is smaller: ¼ or ⅓?
2) Which is larger: ¾ or ⁴⁄₆?
3) How many ⅙s make a whole?
4) How many ⅛s equal one half?
5) What fraction does ⅓ + ⅙ equal?
6) What does ¾ + ⅜ equal?
7) Which is larger: ⅜ + ¼ or ¾ + ⅛?
8) Is ⅔ + ¼ less than one whole?

Directions:

For individual student:

1. Give student one set of fraction bars, one pair of scissors, and appropriate list of questions.

2. When student is finished, he or she should check answers with teacher or be given an Answer Key. Any incorrect answers should be redone using the fraction bars.

For entire class:

1. Divide class up into small groups with each group receiving one set of fraction bars, scissors, and question list.

2. Students work out problems on list as a group, using fraction bars and cutting the bars with scissors when necessary to answer questions.

3. Answers should be recorded. When all groups have finished, the answers should be discussed and checked. Teams receive one point for each correct answer.

MAKE-A-WHOLE

Summary: Students turn over fraction cards hoping to find a pair with the same denominators whose sum is one. For two teams or two players.

Skills: Adding fractions with the same denominators

Supplies: 20 index cards, pen (optional: reusable adhesive tacks).

Preparation: Label the 20 index cards with the following 10 fractions, using each fraction twice: $\frac{1}{2}$, $\frac{1}{3}$, $\frac{2}{3}$, $\frac{1}{4}$, $\frac{2}{4}$, $\frac{3}{4}$, $\frac{1}{5}$, $\frac{2}{5}$, $\frac{3}{5}$, $\frac{4}{5}$.

Directions:

1. Divide class into two teams.

2. Spread out the 20 cards and place them face down in front of the class. They could be placed
   a) on a table
   b) along the chalk tray, or
   c) on the chalkboard using reusable adhesive.

3. A student from the first team goes up to the fraction cards and turns over any two cards, showing them to the rest of the class.

4. If the two cards contain fractions whose sum is one, such as $\frac{1}{4}$ + $\frac{3}{4}$, the student keeps the two cards. If not, the cards are turned face down again and returned to original position.

5. In either case, a new student from the other team comes up and again turns over two cards, keeping cards only when they equal a whole.

6. As the game progresses, students try to remember the positions

of certain face-down fractions, which they might wish to turn over on a later turn.

7. After all cards have been taken, the team with the most cards wins.

FRACTION BEAN POT

Summary: Following the directions of a spinner, students take ½, ⅓, or ¼ of the number of beans currently in the "bean pot." For small groups.

Skills: Fractions as part of a set

Supplies: Two large paper plates or circular pieces of tagboard, brass fastener, paper clip, beans or other counters (10 per student).

Preparation: One plate is to be made into a spinner, and the other plate should be labeled "BEAN POT." For the spinner, first divide circle into eight parts and label as shown in Figure 4-7. Next, attach a brass fastener through an opened paper clip and then through center of circle—loosely enough to allow paper clip to spin freely.

 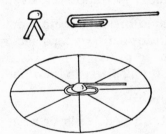

Figure 4-7

Directions:

1. Each student is given 10 beans.

2. In turn, each student spins the spinner. But before each spin, the student must place one bean in the pot.

3. According to the spinner, the student might have to give more beans to the pot, or take some beans back from the pot.

4. If there are six beans in the pot and spinner reads "take ⅓," then student will take two beans from the pot, leaving four. But, if there were only one or two beans in the pot, the student would not take any, since there needs to be at least three beans in the pot in order to take ⅓. Likewise, there must be at least four beans in pot before student can take ¼.

5. If there had been seven or eight beans in the pot and student was to take ⅓, the student would still take two. It would not be until there were nine beans in the pot that student would take three.

6. Students continue, each turn placing one bean in the pot and then spinning the spinner.

7. As students use up their 10 beans they are out of the game. The last student remaining in the game is the winner.

Note: Students should be reminded to always place one bean in the pot at the beginning of their turn.

REDUCE THREE

Summary: A variation of tic-tac-toe where students may only place numbers in squares if they can reduce the fractions formed. For two players.

Skills: Reducing fractions to lowest terms

Supplies: Two index cards, felt-tip pens (two different colors).

Preparation: Prepare a small gameboard similar to Figure 4-8 on an index card. Using a second index card, cut eight small discs, to be used as markers. Label these discs 2, 3, 4, and 5, once in one color and again, using a different color.

Directions:

1. Give each player a set of four discs of one color.

2. As in tic-tac-toe, players take turns placing discs on squares; the object, of course, is to get three in a row.

3. When a disc is placed on a square, a fraction is formed: the number on the disc is the numerator; the number on the square is the denominator.

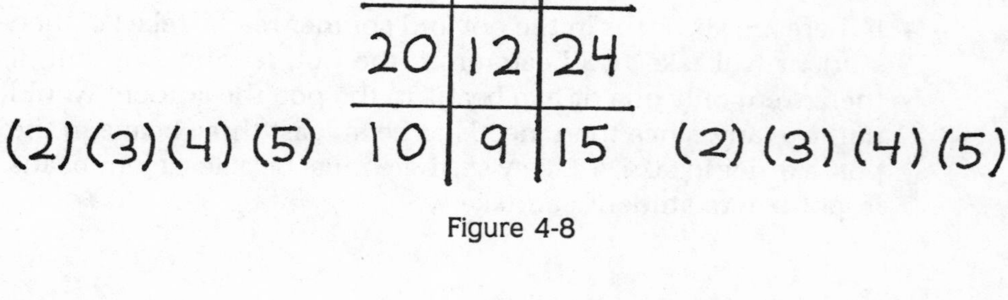

Figure 4-8

4. However, player can only place a disc on a square if the fraction formed can be reduced, and if player can state what it can be reduced to.

5. If, after all discs possible have been placed on board, and if no player yet has three discs in a row, then game continues with players, in turn, moving their discs from occupied squares to empty ones. Of course, all of the above rules must be followed.

Examples: The "3" disc cannot be placed on the "20" square ($\frac{3}{20}$ cannot be reduced), but student may place the "3" disc on the "12" square, stating "$\frac{3}{12}$ equals $\frac{1}{4}$."

FRACTION WAR

Summary: This game for two players, is an interesting variation of the familiar card game "War." Players compare fractions, deciding which is larger.

Skills: Comparing fractions with unlike denominators

Supplies: 12 index cards or tagboard, pen, paper and pencil for each student.

Preparation: Cut out 24 cards from tagboard, or use 12 index cards and cut each in half. Label cards with fractions. Following are 24 suggested fractions. Adjust to your students' abilities:

$$\frac{1}{2}, \quad \frac{1}{2}, \quad \frac{2}{2}, \quad \frac{1}{3}, \quad \frac{2}{3}, \quad \frac{3}{3}, \quad \frac{1}{4}, \quad \frac{2}{4}, \quad \frac{3}{4}, \quad \frac{4}{4}, \quad \frac{1}{5}, \quad \frac{2}{5},$$

$$\frac{3}{5}, \quad \frac{4}{5}, \quad \frac{5}{5}, \quad \frac{2}{6}, \quad \frac{3}{6}, \quad \frac{4}{6}, \quad \frac{5}{6}, \quad \frac{2}{8}, \quad \frac{3}{8}, \quad \frac{4}{8}, \quad \frac{5}{8}, \quad \frac{6}{8},$$

Directions:

1. Divide the 24 cards evenly among the players.

2. Players stack cards face down.

3. Each player turns over his or her top card and places it on the table face up.

4. Player having the larger fraction wins both cards. These two cards are now placed face down at the bottom of the winning player's pile.

5. Game continues with players again turning over their top card. In some cases, it will be very easy for players to decide which

fraction is larger (e.g. ¼ and ½), while at other times, players will have to convert to common denominators (e.g. ⅗ and ⅘).

6. If both fractions are equal, to break the tie players place the next three cards face down on the table and a fourth card face-up. The student whose fourth card has the highest sum wins all 10 cards. (If player had less than 4 cards in a fraction war, then his or her last card would be turned over.)
7. Winner is the player who accumulates all the cards. Or players may set a time limit, and student with the most cards wins.

THE FRACTION MAZE

Summary: Students work independently, adding fractions as they move along through the maze. The object is to find a route with the lowest sum.

Skills: Addition of fractions with unlike denominators

Supplies: Paper, pencils, chalkboard.

Preparation: Fraction maze (Figure 4-9) can be written on the chalkboard, or drawn on paper and duplicated.

Directions:

1. Each student will need paper and pencil and will work independently.
2. The object is to go from "START" to the "END," adding the fractions along the way, and finding which route has the lowest sum.

START $\frac{1}{2}$	$\frac{1}{6}$	$\frac{3}{4}$	$\frac{1}{3}$	$\frac{1}{4}$
$\frac{1}{6}$		$\frac{1}{6}$		$\frac{2}{3}$
$\frac{2}{3}$	$\frac{1}{4}$	$\frac{5}{6}$	$\frac{1}{4}$	$\frac{4}{6}$
$\frac{4}{6}$		$\frac{1}{6}$		$\frac{1}{2}$
$\frac{1}{3}$	$\frac{1}{4}$	$\frac{3}{4}$	$\frac{3}{6}$	$\frac{1}{6}$ END

Figure 4-9

3. Students move from square to square, going to the right or to the left, up or down, but never diagonally.

Note: Students should discover that there are many routes which have the lowest sum, $^{48}/_{12} = 4$.

3-2-1 PASS

Summary: A card game involving geometric figures, in which students develop understanding of fractional values. For four players.

Skills: Addition and comparison of fractions with unlike denominators

Supplies: 20 index cards or cards cut from tagboard; paper and pencil for keeping score.

Preparation: On the 20 cards, draw the following: two cards with angles, three with triangles, four with squares, five with pentagons, and six with hexagons (also write the numerals 1 through 6 as shown in Figure 4-10).

Figure 4-10

Directions:
1. One student, who is the dealer, mixes the cards and gives five cards to each of the four players.
2. After the players have looked at their cards dealer calls out "Pass Three," and all students pass three of their cards (keeping two) to player on their right.
3. Players again look at their five cards. This time the dealer calls "Pass Two," and players pass two cards (keeping three) to the right. They pass once more as dealer calls "Pass One."
4. Scoring involves the fractional parts of each geometric figure on the five cards that the student now has.
5. Suppose a student's five cards contain two squares and three pentagons. Since there are four squares and five pentagons, the student has ¾ of the squares and ⅗ of the pentagons. Score

would be ¾ + ⅗ = ²²⁄₂₀ = 1²⁄₂₀ or 1¹⁄₁₀. Likewise, two angles, one triangle, and two hexagons would equal ½ + ⅓ + ²⁄₆ = ¹⁰⁄₆ = 1⅙.

6. If necessary, to compare scores, students can change their scores to the lowest common denominator of 60. Highest score wins.

Note: After students have developed strategies (which cards to keep and which cards to pass) to obtain the highest score, the game might be changed to allow the student with the lowest score to win.

CATCH THE THIEF

Summary: This game, for two players, is played on an 8 by 8 square board (or checkerboard). Players multiply and divide fractions, then move according to the results.

Skills: Multiplication and division of fractions

Supplies: 20 index cards, tagboard, two different markers, paper and pencil. (Checkerboard and two checkers may be used instead of tagboard and markers.)

Preparation: If checkerboard is not available, draw an 8 × 8 grid on a piece of tagboard. On the 20 index cards, write the folowing problems:

$$\frac{1}{2} \times \frac{1}{2} \qquad \frac{2}{3} \times \frac{3}{4} \qquad \frac{2}{3} \div \frac{2}{5} \qquad \frac{1}{6} \div \frac{2}{3}$$

$$\frac{1}{2} \times \frac{2}{3} \qquad \frac{4}{6} \times \frac{3}{5} \qquad \frac{1}{2} \div \frac{2}{5} \qquad \frac{5}{8} \div \frac{1}{2}$$

$$\frac{1}{2} \times \frac{4}{5} \qquad \frac{8}{10} \times \frac{5}{6} \qquad \frac{1}{3} \div \frac{2}{3} \qquad \frac{3}{4} \div \frac{1}{2}$$

$$\frac{1}{3} \times \frac{6}{8} \qquad \frac{6}{8} \times \frac{4}{5} \qquad \frac{3}{4} \div \frac{3}{5} \qquad \frac{2}{3} \div \frac{8}{9}$$

$$\frac{4}{5} \times \frac{5}{6} \qquad \frac{4}{7} \times \frac{7}{8} \qquad \frac{1}{4} \div \frac{3}{4} \qquad \frac{5}{6} \div \frac{2}{3}$$

Directions:

1. Players place markers on diagonally opposite corners of the grid board. One player is "the detective," the other "the thief."

2. Players take turns picking cards and multiplying or dividing the fractions. The answer must be reduced to lowest terms.

3. After each example is done, players look at the answer. The numerator represents the number of squares the thief may go, while the denominator represents the number of squares the detective may move. On each turn, the thief moves first.

4. To catch the thief, the detective must land in the thief's square by an exact count.

5. Players move from square to square by going up, down, left, right or any combination of these, but never diagonally.

6. Of course, sooner or later, the detective will catch the thief. The idea is to see how many turns it takes for this to be done.

7. Next, the players reverse roles and compare the number of turns taken by the second detective with the number taken by the first.

Note: If answers are properly reduced, the numerators and denominators should contain integers no larger than 5. For example, $\frac{5}{6} \div \frac{2}{3} = \frac{15}{12} = \frac{5}{4}$. (Thief moves 5; detective 4.)

DECIMAL PAC-MATH

Summary: This game is another variation of the video game "Pac-Man." Students add decimal numbers, and these sums control the movement of both Pac-Man and the monsters chasing him.

Skills: Addition of decimals

Supplies and Preparation: See "Pac-Math" in Chapter 2 for diagram and preparation of gameboard. In place of the 12 two-digit addition cards, use these 12 decimal numbers: 1.728, 2.843, 4.337, 21.53, 34.95, 43.54, 53.82, 72.61, 154.6, 213.7, 561.2, and 842.4.

Directions: This game will be played in the same way as "Pac-Math" in Chapter 2, with this exception: First the two decimal numbers are added and checked and their sum looked at. Then Pac-Man will move according to the digit immediately to the left of the decimal point (the units digit) and the monsters will move according to the digit immediately to the right of the decimal point (the tenths digit).

Example: The numbers 154.6 and 53.82 have been picked. They are added to get 20<u>8</u>.<u>4</u>2; and now Pac-Man will move 8 spaces followed by the monsters who may move 4 spaces.

DECIMAL RACETRACK

Summary: Students answer problems involving decimals, and move along a racetrack on the chalkboard by tenths, according to the number of correct answers.

Skills: Decimals, all operations

Supplies: 10 index cards, pen, chalkboard and chalk; paper and pencil for each student.

Preparation: On index cards, prepare problems involving decimals. Have three problems on each card: one easy, one moderate, and one more difficult.
On the chalkboard, draw a decimal racetrack similar to the one below (omit the letters).

Directions:

1. Divide the class into two teams, team A and team B.

2. Have a student from each team come to the chalkboard. Give each student a different card, and one student should write the three problems for their teams on the chalkboard.

3. Meanwhile, the remainder of the students are at their seats, working out the three decimal problems belonging to their team.

4. The teacher will call on three students from each team for the three answers. The teacher should match the level of difficulty of the problem with the ability of student called upon to answer that particular problem.

5. On each round, teams receive scores as follows:

    1 correct = .4   2 correct = .7   3 correct = 1.0

For every tenth in the team's score, a letter is placed on the racetrack. Figure 4–11 shows team A with 1.1 points and team B with .7 points.

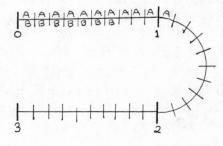

Figure 4-11

6. More rounds are played and the first team to reach "FINISH" (3.0) wins. Depending on students' abilities, between three and five rounds are usually needed.

EQUIVALENCE LOTTO

Summary: Students cover squares on a lotto board by finding equivalent forms of fractions, decimals, and percents.

Skills: Fractions, decimals, percents

Suppies: Tagboard, pens, markers, paper and pencil for each student, chalkboard.

Preparation: Cut out squares to be made into 4 by 4 lotto boards, each one different. Two are shown in Figure 4-12. Each fractional number should appear in many equivalent forms: $3/5 = 6/10 = .6 = .60 = 60\%$. More than one form of a fraction might appear on the same board.
Students will work in small groups of two or three, and one lotto board will be needed for each group.

Directions:

1. Each group of two or three students should have one lotto board, 16 markers, and a paper and pencil for each student.

2. The teacher, or a leader, should call upon each group in turn, to suggest a fraction, decimal or percent, which is then written on the chalkboard.

3. Students, using paper and pencil, search for equivalent forms and cover all they can find on their lotto boards with markers.

4. If, for example, $6/16$ was suggested, teams might reduce it to $3/8$ and then convert it to a decimal ($.37\frac{1}{2}$) and to a percent ($37\frac{1}{2}\%$). Any of these found on their lotto board would be covered.

$87\frac{1}{2}\%$	$\frac{3}{5}$	$.5$	$\frac{1}{2}$
$.60$	$\frac{1}{100}$	$\frac{1}{3}$	$.05$
$\frac{1}{6}$	$.7$	60%	$\frac{7}{8}$
$\frac{2}{6}$	25%	$.01$	$.16\frac{2}{3}$

$.50$	90%	$\frac{1}{4}$	$.9$
$\frac{4}{12}$	$.03$	$\frac{1}{12}$	50%
$12\frac{1}{2}\%$	$\frac{5}{10}$	$.25$	$.70$
$\frac{5}{100}$	$8\frac{1}{3}\%$	$\frac{1}{8}$	1%

Figure 4-12

5. After the first group covers all 16 squares on their lotto board, these numbers are checked against the numbers on chalkboard. If all is correct, that team wins.

Note: The fairness of this game will depend upon establishing an order in which teams will be picked to suggest numbers. While teams might all wish to be the first, there is an advantage to being called later on when team might need only one or two more numbers.

THE JUNGLE GAME

Summary: A complex and challenging game for two players, where pieces representing animals capture other animals smaller than themselves. The score is originally a fraction (based on the comparative sizes of the two animals), then converted to a decimal. This game is suitable for students who already know how to change fractional numbers to their decimal equivalents.

A variation is later mentioned using only multiplication and addition of integers, to be used by students not yet able to change fractions to decimals.

Skills: Conversion of fractions to decimals

Supplies: Tagboard, red and blue felt-tip pens; paper and pencil for keeping score.

Preparation: Draw a 7 by 7 gameboard, labeled with 16 animals as shown below. Cut 16 circular pieces out of tagboard and label with animal's initial and animal's numerical value (Figure 4-13). Make one set of eight pieces in red, the other eight pieces in blue. Animals' values are as follows:

Elephant—8	Fox—4
Lion—7	Dog—3
Tiger—6	Cat—2
Wolf—5	Mouse—1

The back of each piece is to be labeled using the same letter and number as the original side, but preceded by the letter S. Examples: the back of D-3 would be labeled SD-3; the back of F-4 would be labeled SF-4, etc. (These "flip" sides will be used later in the game when T-6, the tiger, might be converted to ST-6, super-tiger.)

Directions:

1. After gameboard is set up as shown, each player, in turn, may

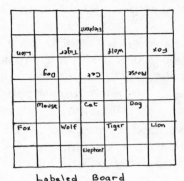

Labeled Board

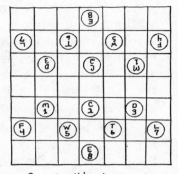

Board with playing pieces

Figure 4-13

move any one of his or her pieces one space in any orthogonal direction—up, down, left or right.

2. However, a player may not place a piece in a square already occupied by his or her own piece.

3. A capture is achieved (as in chess) by landing a piece on a square occupied by an opponent's piece, but, each piece may capture an opponent's piece, only if the opponent's piece represents a smaller animal. *Example:* a lion (7) may capture a fox (4), but it may not capture an elephant (8). Also, an animal cannot capture another animal of equal value. Captured pieces are taken off the board.

4. To score such a capture, make a proper fraction using the two values, and then convert it to a decimal. If a lion captured a fox the score would be $\frac{4}{7}$, and when changed into a decimal, this would become .571 (use up to three decimal places).

5. There are two special rules, both involving the mouse:

   a) Since the elephant is the largest animal, none of the other pieces can capture it; however, a mouse may land in a square occupied by an opponent's elephant and scare it off the board. (Score would be $\frac{1}{8}$ = .125) Moreover, the elephant is not permitted to capture the mouse.

   b) The only animal allowed to capture the mouse is the cat. (Score is $\frac{1}{2}$ = .500)

   Aside from these rules, animals will only capture animals smaller than themselves.

6. As players achieve certain scores, they may turn over any one of their pieces. When a piece gets flipped over to its "S" side it has a

special privilege: it can now move one space diagonally in addition to orthogonal moves (of course, only one move per turn). As soon as a player's total score is 1.000 or higher, a player may flip over any one piece; at 2.000 or over, player may flip over a second piece and at 3.000, he or she may flip over a third piece. When L-7 (lion) gets flipped over it becomes SL-7 (super-lion), and it may now move one space in any direction.

7. The game ends when the once overcrowded jungle of 16 animals is no longer crowded and only five animals remain. At that point, the players leave these five pieces on the board and look at their scores. The player with the highest score wins.

Note: The player with the most captures does not necessarily win the game. It is the comparative value between the captured and capturing pieces that counts. One capture of wolf by tiger ($5/6$ = .833) is larger than the sum of these two captures: cat by lion ($2/7$ = .285) plus dog by elephant ($3/8$ = .375).

Variation: Instead of using fractions or decimals, this variation can use multiplication and addition of integers. The scoring is computed as follows: Each time a capture is made, students multiply the value of the two pieces. Students will turn pieces over to their "S" sides when their total score is equal to or larger than 50, 100, 150, 200.

In the multiplication variation, when a tiger (6) captures a fox (4), the score will be 24. If, on the player's next turn, the lion (7) captures the opponent's wolf (5), the player receives 35 points and has a total score of 59. This player may now flip any piece over to its "S" side. All other rules remain the same.

CHAPTER 5

Quick and Easy I—
A Collection of Games
That Require Little Preparation—
Just Pencil, Paper, Chalkboard
and Chalk

The 28 games in this chapter are quite varied; some are competitive, some are cooperative, and all involve the entire class. Some require a small amount of preparation, others none at all. Each game in this chapter is designed to sharpen a specific skill. While the following chapter will contain more Quick and Easy games, those will be adaptable to different needs.

Games like "Choose Your Place" will help students understand the importance of place value, while "Fish Pond" offers a fun way to practice addition combinations. "Palindromes" will have your class adding three- and four-digit numbers as they experiment with number patterns. "School Pool" or the "Ultimate Multiplication Contest" will challenge students to try different multiplication combinations, in a search for the largest possible solution.

Also included are games that specifically deal with money, measurement, fractions, decimals, and percents. If you do not find activities here that meet the exact needs of your students, try Chapter 6, with its more adaptable games. Be sure to consult the Subject Index which will refer you to various games and activities throughout this book.

NUMBER OF THE DAY

Writing and understanding the numerals 1 to 10

Preparation: None.

Directions:

This activity takes 10 days. Each day draw a large numeral on the board. "1" on the first day, "2" on the second day, continuing up until "10" on the tenth day. The children copy this numeral on a sheet of paper. The idea is to find things in the classroom of which there are exactly that number. A list may be written and pictures drawn.

Examples: for the sixth day—six boys with blue shirts, six lights on the ceiling, six books on the teacher's desk, six flowers on the bulletin board. Check to make sure appropriate items are in the classroom each day.

COPY CAT

Silent counting

Preparation: None.

Directions:

You might wish to start the game by being the leader yourself. Later, select students to be the leader, changing leaders frequently.

The leader will repeat an action a certain number of times, such as jumping or clapping hands. The class silently counts while watching the leader. Then the leader may call upon one student, a group of students, or the whole class, to be the copy cats and repeat this action the correct number of times.

When the class has no difficulty copying one action, have the leader do two activities, such as tapping shoulders six times, then bending over four times.

This game may also be played as a competitive game with elimination as in "Simple Simon."

TWENTY

Writing numerals to 20

Preparation: None.

Directions:

Divide students into two teams to play this game. Teams alternate in sending a student to the chalkboard. Starting with the number 1, students, in turn, will write the next consecutive number or the next two consecutive numbers on the chalkboard. Thus, the first student might write 1, or 1 and 2, but no more than these two numbers. Likewise, if the numbers from 1 to 12 were already on the board, the next student might write only 13, or write 13 and 14. The team that writes 20 wins.

GET IN ORDER

Ordering two-digit numbers

Preparation: None.

Directions:

Each student will need a piece of paper. Ask students to write any two-digit number. Numerals should be written large enough to be seen by the whole class. Most likely, numbers will be different with only a few repeated.

First, you might ask students to stand up (holding their papers) when their numbers are called: "All numbers less than 27," "Numbers more than 35," "Only numbers more than 20 but less than 50."

Next, students will be involved in ordering. Call up a group of six to 10 students (with their numbers) and ask them to arrange themselves in numerical order. Repeat with other groups, until all the students have been called up at least one time. Finally, have the entire class get up and attempt to line up in order.

CHOOSE YOUR PLACE

Place value—3, 4, or 5 places

Preparation: You will need nine cards with the digits 1 to 9 written on them.

Directions:

Draw three to five consecutive blank squares on the chalkboard, which students should copy at their seats. A few students might go up to the chalkboard. Suppose you draw four squares; you would then pick four cards, one at a time, calling out the digits to the class.

Each time a digit is picked, each student is to write that digit down in one of the blank squares. By the time the last digit is called, there should be only one place left, and that will complete the number. Then, students should each read their number. The student having the largest number wins.

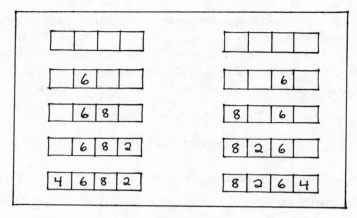

Figure 5-1

NUMBER DETECTIVE

Numbers from 1 to 100, logic

Preparation: None.

Directions:

Select a student to come up to the front of the room. This student writes a numeral between 1 and 100 on a piece of paper and gives it to the teacher.

The remainder of the class will be the "number detectives," and will try to guess the number by asking this student a series of questions. All questions should be the type that can be answered by "YES" or "NO." A game might go like this:

"Is your number greater than 60?"	"No"
"Is it less than 40?"	"Yes"
"Is the ten's digit 3?"	"No"
"Is it a two-digit number?"	"Yes"
"Is it less than 20?"	"No"

If you wish to make this a competitive game between two teams, add this rule: students from one team may continue asking questions as long as the answer is "YES." As soon as a team receives a "NO" answer, the other team asks the questions, until, of course, this team receives a negative response.

HEADBAND

Numbers from 1 to 100

Preparation: Prepare a few long strips of paper that can be made into headbands. On each headband write a numeral between 1 and 100.

Directions:

Place a headband on a student volunteer so that this student does not see the written numeral. This game now becomes similar to *Number Detective*, but is reversed.

Your class should all be able to see the headband numeral, except, of course, for the student wearing the headband. This student may now ask each member of your class a "YES" or "NO" question, as in the preceding game.

FISH POND

Addition and/or subtraction

Preparation: None.

Directions:

Draw the outline of 20 fish on the chalkboard, each with an addition or subtraction combination written on it. Call a student to go fishing for all the fish with a certain sum or difference. For example, a student might be asked to go fishing for all the "8" fish. This student would then circle all the fish that he or she thinks have the sum (or difference) of 8, which your class should then check. If correct, the student erases all these circled fish. Another student is then called upon for a different sum. This continues until the fish pond is empty.

(This game is a Quick and Easy variation of *Let's Go Fishing* in Chapter 2.)

LADY BUGS

Addition: a) using numbers 1 to 5, or b) using numbers 6 to 10

Preparation: Depending on your students' needs prepare 10 blank cards and a) label with numerals 1 to 5, using each numeral twice; or b) label with numerals from 6 to 10, using each numeral twice. These 10 cards should be placed face down on a table in front of the classroom.

Directions:

Divide your class into teams. Each team will try to be first to complete their lady bug. A completed lady bug has 15 parts in addition to its body: six legs, four spots, one head, two eyes, and two antennae.

Each team starts by drawing a body of a lady bug on the chalkboard. From now on, students (teams alternate) come up, one at a time, to the front of the room and pick two cards, which are shown to

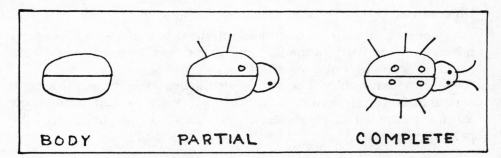

Figure 5-2

the class. Depending on the sum of these two numbers, a part may be added to the team's lady bug, if it is needed.

According to the chart below, a) 2 + 4, or b) 7 + 9, means the student may draw one leg on the lady bug. Cards are replaced, (face down), and mixed before the next student comes up. Please note that the head must be drawn before an eye or an antenna may be used.

Before beginning the game, draw a diagram of a completed lady bug and the appropriate portion, (a) or (b), of the following chart on the chalkboard.

(a) 1 - 5 SUM	PARTS	(b) 6 - 10 SUM
4	Eye	14
5	Spot	15
6	Leg	16
7	Head	17
8	Antenna	18
2 or 10	Any part needed	12 or 20
3 or 9	No parts	13 or 19

PINE TREES

Addition or subtraction

Preparation: None.

Directions:

Draw a few pine trees on the chalkboard and many pine cones. The pine cones are labeled with addition (or subtraction) combinations, while the pine trees are labeled with the sums (or differences). Students come up to the chalkboard and draw lines from the pine cones to the corresponding trees.

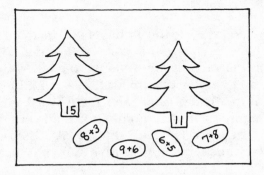

Figure 5-3

QUICK AND EASY TIC-TAC-TEN

Addition of three numbers

Preparation: None.

Directions:

Draw a 3 by 3 tic-tac-toe grid on the chalkboard with a list of numbers 1 to 5 for each team.

Teams take turns placing numbers in the grid, crossing out numbers as used. The object is to place a third number so that the sum of all three numbers in a row is 10. Please note that once a number is crossed out it cannot be used again by that team.

(For more details on the strategy of this game, see "Tic-Tac-Ten" in Chapter 2.)

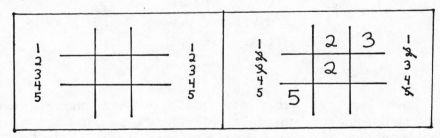

Figure 5-4

MATH ALPHABET

Addition using one- and two-digit numbers

Preparation: You might wish to find the math alphabet sum for a few
words in advance.

Directions:

On the chalkboard write the alphabet and its math equivalent: A (1), B (2), C (3), D (4), E (5), ... , up to X (24), Y (25), and Z (26).

The idea of this game is for students to guess a word (or a name) when given the math alphabet sum for that word. Clues must be given. For example, the math alphabet sum for a word is 39. "It has four letters." "There are many in this room." "It is made out of wood." "We do our work on it." The word is DESK: D (4), E (5), S (19), and K (11).

How about a three-letter word with a sum of 11, useful when tired?

PALINDROMES

Addition of three- and four-digit numbers

Preparation: None.

Directions:

First explain to your class that a palindrome is a number or word that is the same backward as forward, such as 4664 or 12721.

Next, show them an interesting mathematical experiment: if you take a number, reverse the digits and add them to the original number, then take that sum and reverse the digits and add, sooner or later, you will get a palindrome. Here are a few examples using three-digit numbers:

235	763	361	482
532	367	163	284
767	1130	524	766
	0311	425	667
	1441	949	1433
			3341
			4774

Have your students try this method for making palindromes with some of the following numbers: 152, 361, 673, 249, and 957. These numbers should take two or three additions to become palindromes. Challenge your students to find numbers that take more than three reversals and additions to make a palindrome.

LET'S GO SHOPPING

Money

Preparation: None.

Directions:

On the chalkboard draw pictures (with prices) of objects that students might wish to buy. It might be fun to have students help with these diagrams. Next, draw the coins, enough to buy some of the objects, but not all.

Call students to the chalkboard, one at a time. The student circles an object that he or she wishes to buy, then crosses out coins necessary for the purchase. This shopping trip can continue (as in real life) until there is not enough money left to make any more purchases.

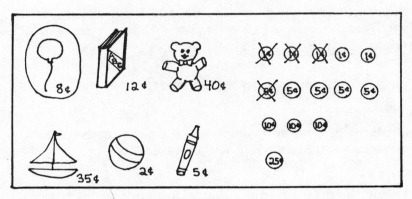

Figure 5-5

MONEY BAG

Money

Preparation: Real coins will be most useful. If not available, cut small
   circles out of heavy paper and label them accordingly.

Directions:

Place a series of coins in a small bag or envelope. Give your class two clues—the number of coins, and the total amount. (Example: four coins with the total of 37 cents) Students may wish to use paper and pencil, and as soon as the students know the answer they should write it down and then raise their hands.

After most of the class has their hands raised, call on one of the students for his or her answer. Ask the rest of the class how many students had the same answer. Have a a student look in the money bag to check answers. Change coins in the money bag (or let a student do it) and repeat.

DRAW IT

Measurement

Preparation: Students will need rulers.

Directions:

Ask your students to draw some objects correctly, concentrating on their sizes and shapes. This should be done from memory, without looking at the actual items.

Suggest objects that have a standard size, such as a penny, a dollar bill, a new crayon, a drinking straw, etc.

After several objects have been drawn, ask the students to measure the lengths of the objects in each drawing. Compare student estimates to the length of the actual item.

MEASUREMENT DETECTIVE

Measurement

Preparation: Prior to the game, measure a variety of the objects in your classroom, and record these measurements.

Directions:

Divide the class up into many teams, and give each team a card containing several sets of measurements. A list might include measurements in centimeters (e.g., 54 cm × 65 cm; 11 cm × 12 cm × 18 cm) or inches (11 in. × 14 in.; 26 in. × 52 in. × 29 in.). Your students will need the appropriate rulers.

Students move around the room measuring objects, as they attempt to find the exact items described. The items might be desks, doors, pictures, boxes, bulletin boards, books, or cabinets.

MULTI-HOPSCOTCH

Multiplication

Preparation: Immediately before the game, with chalk, draw a hopscotch diagram on the floor.

Directions:

Write a few numbers on the chalkboard to be used as multipliers (e.g., 4, 5, 6), then select a student to choose one of the numbers as a multiplier. He or she jumps (or hops) into a hopscotch box and proceeds to multiply each of the numbers in the boxes by this chosen

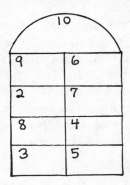

Figure 5-6

multiplier. The student says "3 × 4 = 12." The remainder of the class listens, and if the multiplication is correct, calls out, "Jump" and the student will jump into the next box. This continues until the student calls out an incorrect product at which time your class should call out, "Stop."

After a student successfully calls out all nine products correctly, that number is crossed out on the chalkboard and another multiplier is chosen. Hopscotch continues, with students taking turns, until all the multipliers have been eliminated.

SCHOOL POOL

Multiplication

Preparation: None.

Directions:

Draw a pool (billiards) diagram on the chalkboard, similar to the one below.

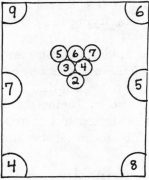

Figure 5-7

Have students imagine that this is the top of a pool table; the circles in the center are the balls and the semicircles on the sides represent the pockets. Each ball goes into a different pocket. The score is the product of the number on the ball and the number of the pocket.

For example, a "3" ball in the "7" pocket contributes 21 points towards the score. After trying a few more examples, ask your class, "If you could only use each pocket once, how would you play this game to get the highest score?" Allow students to work individually, then compare total scores.

THE ULTIMATE MULTIPLICATION PROBLEM

Multiplication using two-digit or three-digit or five-digit numbers

Preparation: None.

Directions:

Your class should enjoy these problems. Let them start with the first and then, if they are able, proceed in order. A few days after the class has discussed solutions to the second problem introduce the "ultimate" problem.

a) An interesting multiplication problem: Using the digits 1, 2, 3, and 4, form two two-digit numbers, and multiply them. Each digit should be used only once. Try to find the combination that gives the greatest product. Is it 12×34, 13×24, 32×14, or an entirely different arrangement?

b) A challenging multiplication problem: Using six digits, 1, 2, 3, 4, 5, and 6, form two three-digit numbers that have the greatest product. (Problem can now be expanded to eight digits.)

c) The ultimate multiplication problem: This time have your class use all 10 digits, 0, 1, 2, 3, 4, 5, 6, 7, 8, and 9. Using each digit only once, form two five-digit numbers, trying to find the combination that gives the greatest product.

There are over three million combinations, but your class will not need to try them all if they use insights gained from the solutions to the three previous problems. It turns out that the solution can be found by continuing the pattern found in the last example.

In case you might be interested in working on these problems yourself, the answers will be found at the end of this chapter.

DIVISIBILITY

Division

Preparation: None.

Directions:

Students can work individually or in small groups. Write a large number on the chalkboard, three- to five-digits. Within a given time limit, students try to find as many numbers as possible that divide into it without leaving a remainder. For example, 24,678 is divisible by 2, 3, 6, and 9. Have your class try some of these large numbers: 336; 4,305; 20,250; 46,656.

FRACTION ACTION

Fractions

Preparation: None.

Directions:

All students will need paper and pencil. Ask each student to draw a picture of a fraction. Tell them the idea is to see how many different kinds of fractions the class can draw.

Then write a fraction on the chalkboard, such as "¼." Ask all students who have drawn pictures of ¼ to come to the front of the room and show the class. Discuss each one. Continue with common fractions such as ½, ¾, ⅓. Then ask students who have other fractions to come up and explain their pictures to the class.

Each time a student is the only one to illustrate a particular fraction, compliment him or her and tape this fraction to the chalkboard. Count the number of different fractions that were drawn.

Play the game again: this time there will probably be many more "unique" fractions.

MUSICAL FRACTIONS

Addition of fractions

Preparation: Bring in a book or a sheet of written music.

Directions:

Show the class musical notes—a whole note, a half note, a quarter note, and an eighth note. Leave these sample notes on the chalkboard. Copy some notes from a piece of music; have your class add up these fractions.

You might also discuss the fraction at the beginning of the first line of music, and the total of the notes in each bar. Can your students find the connection?

FRACTION SCORE

Addition of fractions

Preparation: None.

Directions:

This game can be played by two teams or two players. Draw a diagram consisting of 16 circles on the chalkboard, arranged as shown below. Leave the circles blank.

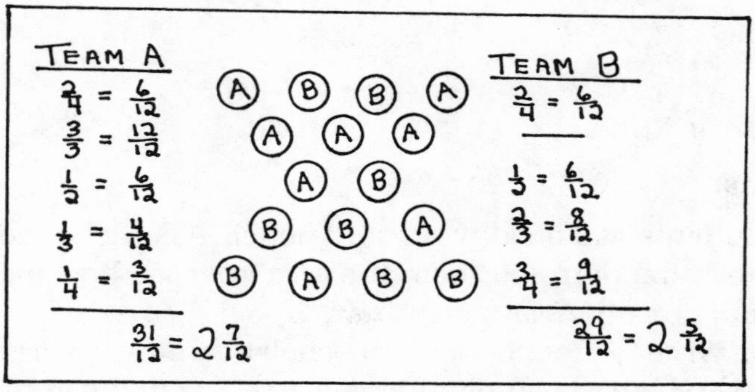

Figure 5-8

Players take turns filling in the circles, using initials. Score is found by listing the fractional part of each line containing the team's initial, and then adding. In the example above, Team A has two out of four or $\frac{2}{4}$ of the top line; three out of three or $\frac{3}{3}$ for the second line; one out of two, or $\frac{1}{2}$ of the third line; one out of three or $\frac{1}{3}$ of the fourth line; and $\frac{1}{4}$ of the fifth line. Team A has a total score of 2 and $\frac{7}{12}$, and wins this game. Note that the sum of Team A's and Team B's score must always equal 5.

Before long, students will know which are the best circles to fill in first, and this will result in a constant tie score. Try changing the diagram around and playing for the lowest score.

FRACTION DIAGRAMS

Multiplication of fractions

Preparation: None.

Directions:

Divide the class into three or four teams. First illustrate on the chalkboard how to use fraction diagrams to solve problems involving multiplication of fractions.

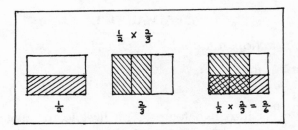

Figure 5-9

Send one student from each team to the chalkboard and announce a fraction multiplication problem. All three or four teams will try to solve this same problem by using diagrams. The first team done gets five points, second team done gets four points, third team gets three points and if there is a fourth team they would get two points.

Send a new set of students to the chalkboard and continue. After several rounds, total up the scores. Fractions should be simple and easy to illustrate. For example: $\frac{1}{3} \times \frac{1}{2}$, $\frac{1}{4} \times \frac{1}{3}$, $\frac{1}{2} \times \frac{3}{5}$, $\frac{3}{4} \times \frac{1}{2}$.

DECIMAL TEAM—FRACTION TEAM

Decimals and fractions

Preparation: None.

Directions:
Allow each student to decide which team they wish to join. Decimal team will use only decimals; fraction team will use only fractions. Present the following matched set of problems:

$\frac{1}{2} + \frac{4}{5}$	.5 + .8
$\frac{3}{8} \div \frac{1}{4}$	.375 ÷ .25
$\frac{3}{4} - \frac{3}{10}$	.75 − .3
$\frac{1}{4} \times \frac{2}{5}$	.25 × .4

Students do all four questions belonging to their team. Note which teams the first 10 students to finish were from. Is either method faster? How many students from each team got all four answers correct? Is either method easier for addition, for subtraction, for multiplication, for division?

DECIMAL DUEL

Addition, subtraction, multiplication and division of decimals

Preparation: None

Directions:

Divide class into two teams. The first team suggests a decimal number (for example, 38.1). The second team suggests another decimal number (for example, 2.58). Immediately, both teams add, subtract, multiply, and divide these two numbers. (In division, the larger number is divided by the smaller.)

The team that finishes first will write the four answers on the chalkboard. The other team will check these answers. If the first team is correct in all four answers, that teams gets one point. If any of the answers are incorrect, then the other team receives the point.

A few rounds should be played, and first team to get three points is the winner.

THE $100 SALE

Percentages

Preparation: None.

Directions:

Write the following "advertisement" on the chalkboard. Illustrate if you wish.

YEAR-END GOING-OUT-OF-BUSINESS-SALE		
10% OFF	15% OFF	20% OFF
Electronic games ...$18	Tennis Racket ...$20	Record Album ...$13
Jacket ...$35	Camera ...$40	Radio ...$28

Break the class up into groups. Five items can be purchased whose reduced sale prices total exactly $100. Members of each group will need to cooperate with each other if they are to be the first group to find the solution.

ANSWERS FOR THE ULTIMATE MULTIPLICATION CONTEST

a) 41×32

b) 631×542 (8531×7642)

c) 96420×87531

CHAPTER 6

Quick and Easy II—
Adaptable Games That Can Be Used
to Meet the Individual Needs
of Your Classroom

As the teacher, you know your class best. You know what subject matter should be included in a game. You know how your class would react to an action game, as opposed to a quiet one. You know the best time to play games. Is it early in the school day, just before or after more formal math activities, or at the end of the day? Would games serve to motivate your students toward your planned math lesson, or should the games be used as the math lesson?

As you read each of the 21 games and activities in this chapter, picture how it will fit into your classroom. How does it meet the needs of your students? Most games are excellent for reviewing basic math facts, but some are more suitable for more complex questions and longer problems. Three of the activities here stress the importance of relating math in the classroom to math in the real world.

You will probably want to use games for a variety of situations, and to fit a variety of needs. If you do not find all that you need here, perhaps you might want some permanent premade games found in earlier chapters which can be used independently by groups of students. You might also want to include a historical or puzzle approach to mathematics, found in the later chapters.

CONCENTRATION

Pairs of cards are matched. Many applications are suggested.

Preparation: On 20 small cards write numbers (or objects) so that there are 10 matching pairs. Here are some possibilities:

Ideas	Examples
Numerals	"6" will match with "6"
Counting	6 objects will match with 6 objects
Numbers and Counting	"6" will match with 6 objects
Addition	Combinations with same sum will match; 6 + 2 will match 3 + 5
Geometry	Square will match with square
Decimals and Fractions	.6 will match with $\frac{6}{10}$

Directions:

Cards are turned face down and spread out. Two cards are looked at on each turn. If they are a match they are removed. If not, they are returned face down to original position. Game continues until all cards have been matched.

CONNECT A PAIR

Using chalk, students connect corresponding pairs.

Preparation: This game should be prepared in advance or done spontaneously on the chalkboard.

Directions:

In a rectangle drawn on the chalkboard, write a variety of questions and their solutions; each should be circled. In the illustration below, addition and subtraction combinations are shown, but any combinations could be used.

Students in turn connect a pair of corresponding numbers, drawing lines carefully so they do not cross or touch other lines or circles. Continue until all pairs are connected.

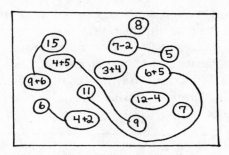

Figure 6-1

If you wish to make the game competitive, choose teams and give a point for each correct match. Do not give a point if a line touches others; leave it on the board and continue.

MATH WIZ

Students ask questions of a classmate who will attempt to answer seven questions in a row.

Preparation: None.

Directions:

A student volunteer comes up to the front of the room. The remainder of the class takes turns asking math questions. Each time the student volunteer gets the correct answer, he or she gets one of the letters of M-A-T-H W-I-Z. If seven questions are answered, this student receives all seven letters and becomes a "MATH WIZ." The MATH WIZ then retires, and a new volunteer is chosen.

A student may only ask a question to which he or she knows the answer. Each student should be given only one turn to ask questions. If a student asks a question that is not answered, this student comes up to the front of the room (replacing the previous student volunteer) and now attempts to become a MATH WIZ.

PLACE YOUR DIGITS

The object is to place digits to obtain the largest final result. Each digit must be placed independently, before knowing what the next digit will be.

Preparation: You will need 10 cards; labeled with the numerals 0 to 9.

Directions:

When you first introduce this game to your class, start with addition. Later you may use other operations familiar to your class, such as subtraction or multiplication.

Draw four to six rectangles, each to accommodate one digit, in any format you wish. Figure 6-2 shows a form for addition of three-digit numbers, and two different forms for multiplication. Have all the students copy the chosen format on a sheet of paper, with the rectangles left blank.

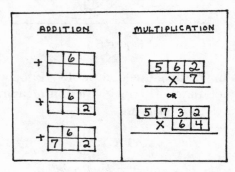

Figure 6-2

After the 10 cards have been mixed select card and read the digit to your class. Each student must place this digit in one of the blank rectangles. Continue selecting cards, one at a time, as students fill in the rectangles. After all the rectangles are filled, each student should work out his or her example. Since each student has worked independently, the digit arrangements and thus the solutions should be different. Students with the largest final results win.

ACTION BASEBALL

Students answer questions and move from base to base as in the game of baseball.

Preparation: In advance, have each of your students prepare three questions based on your recent math classwork: one easy, one moderate, and one difficult. Collect and arrange these into the three categories. You might wish to supplement these questions with a few of your own.

Directions:

Prepare four bases around your classroom: home plate, first, second, and third. Break the class up into two teams. The game will be played like baseball, but when students come up "to bat," allow them to decide if they want to try an easy, a moderate, or a difficult question. The easy questions are worth a single base hit, in which all students on base will move one base, the moderate questions are worth a double, and the difficult questions are a triple. An incorrect answer will, of course, be an out.

To introduce an element of luck into the game, prepare some cards labeled "Automatic Out" and mix them into the three different groups of questions. The game could end after a certain number of innings, or after a certain time limit.

MATHEMATICS IN PRINT

Students bring in printed materials that have mathematical applications.

Preparation: None.

Directions:

Select a volunteer each day to bring in some printed material related to mathematics. It could be an advertisement from the newspaper, a list of sports scores, a recipe, a panel from a box of cereal (nutrition information), a menu, or any other similar item.

Each day use what has been brought in and make up an appropriate problem based on the written information. Even better, have the student bringing in the material make up a problem at home, perhaps with some family help.

The objective here is to have your students see real-life applications of the mathematics they are learning in school.

NUMBER SEARCH

After students solve math problems, they search for the answer in a number grid.

Preparation: In advance, work out a number of problems and place their answers in a grid measuring 4 by 4 or larger. Fill in remaining squares with random digits.

Directions:

Place the math problems and completed number grid on the chalkboard. Have the students copy the grid, solve the problems, and find these answers on the number grid.

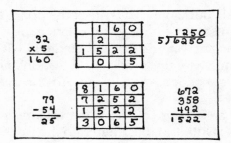

Figure 6-3

Now you can challenge your students to make up their own number grids. On subsequent days these will be shared with the class.

STAND UP/SIT DOWN

A statement relating to math is called out. Students stand up it they agree, sit down if they disagree.

Preparation: None.

Directions:

A student volunteer stands up and calls out a mathematical statement. This statement might be true or it might be false. It could be in the form of an equation. "8 + 6 = 15," or an inequality, "7 × 5 is larger than 8 × 4," or a definition, "a rectangle is a shape with three sides," or a statement based on any other information that has been discussed in class, such as "a centimeter is smaller than an inch."

Each student decides whether the statement is incorrect or incorrect. If a student believes the statement to be correct, he or she should immediately stand up. On the other hand, a student who believes that the statement is incorrect should remain sitting.

If the statement was true, then one of the standing students is chosen to be the new leader; if the statement was false, one of the sitting students would be chosen. With the new leader standing and the rest of the class again sitting down, the leader now calls out another mathematical statement and the game continues.

VOLLEYBALL MATH

In this elimination game, players must answer questions in a game resembling volleyball.

Preparation: For this game, an area of the classroom should be cleared. An object suitable for tossing, such as a ball, will be needed.

Directions:

Prior to playing the game, write on the chalkboard a list of math combinations that your class needs to practice. The class is now to be divided into two teams, who will stand facing each other as in volleyball; a net is not needed, however.

Give one player an object suitable for throwing (a ball, a bean bag, and so on). This player calls out one of the problems on the chalkboard, then throws the ball to a player on the other team. That player must immediately answer the question. Next, a player from the second team calls out a question and throws the ball to a player on the first team.

This continues as players are eliminated. A player will be out if he or she does not answer the question correctly, or if the player drops the ball. (Wild throws do not count.)

Each time a combination is called out, place a check next to it. After the same combination has been used five times, it should be erased. Students should be motivated to listen to all answers since the questions will be repeated.

The team with the most players left after a given time wins.

RECALL

Information is placed on the chalkboard, then erased, and students try to recall the information.

Preparation: None.

Directions:

Information is written all over the blackboard in a random order and arrangement. Here are some examples: 8 to 12 addition or multiplication facts; 6 geometry shapes or definitions; 5 to 10 equivalent fractions. Two ideas are shown below. Adjust students' abilities.

$$3 \times 5 = 15$$

$$2 \times 3 = 6 \qquad \frac{1}{2} = \frac{2}{4} \qquad .05 = \frac{5}{100}$$

$$5 \times 4 = 20 \qquad \frac{1}{3} = \frac{2}{6}$$

$$2 \times 4 = 8$$

$$3 \times 4 = 12 \qquad .6 = \frac{6}{10} \qquad 20\% = \frac{1}{5}$$

After one minute the board is erased and students write down all the facts they can recall.

In a competitive game there would be a large number of facts written on chalkboard and the student with the longest correct list would be the winner. In a noncompetitive game the class would pool information together and try to reconstruct every fact that had been placed on the board.

MISSING DIGITS

Examples, together with answers, are placed on the chalkboard, however some digits are missing.

Preparation: Take examples familiar to your class and omit a few digits.

Directions:

Place examples on the chalkboard. Start with easy examples that gradually get more difficult. Can your students find the missing digits?

$$
\begin{array}{r} 8 \\ -\ * \\ \hline 3 \end{array}
\qquad
\begin{array}{r} *3 \\ +5* \\ \hline 95 \end{array}
\qquad
\begin{array}{r} *2 \\ \times\ \ 4 \\ \hline 128 \end{array}
\qquad
\begin{array}{r} 47 \\ \times\ \ * \\ \hline 94 \end{array}
\qquad
\begin{array}{r} *56 \\ -6*1 \\ \hline 23* \end{array}
$$

More complex problems would include two- and three-digit addition, multiplication, division, and even fractions.

$$
\begin{array}{r} 346 \\ 2*7 \\ 483 \\ 52* \\ \hline 1*07 \end{array}
\qquad
\begin{array}{r} 53* \\ \times\ *6 \\ \hline 3192 \\ *128 \\ \hline 244*2 \end{array}
\qquad
\begin{array}{r} 5* \\ *2\,\overline{)1632} \\ 160 \\ \hline 3* \\ 32 \\ \hline \end{array}
$$

$$\frac{1}{8} + \frac{*}{8} = \frac{6}{8} = \frac{3}{*}$$

$$\frac{3}{*} \times \frac{*}{6} = \frac{6}{24} = \frac{*}{4}$$

FIVE IN ORDER

Combinations are placed on the chalkboard; students rearrange them in numerical order.

Preparation: None.

Directions:

Place five combinations on the chalkboard in random order. Have your students try to copy them, not in the order given, but in numerical order.

GIVEN	ORDERED	GIVEN	ORDERED
7+8	5+6	8×4	5×6
6+7	4+8	5×6	8×4
5+6	6+7	8×6	7×5
8+6	8+6	7×5	6×7
4+8	7+8	6×7	8×6

As soon as the students are finished have them raise their hands. After all your students are done, send those students who finished first to the board and explain their ordered lists.

HORSERACE

Chalk figures of horses fill up a race track as students answer questions.

Preparation: None.

Directions:

Draw a race track on the chalkboard. Below are two possible diagrams—the first for two teams; the second more suitable for three to five teams.

Divide your class into two or more teams, by rows, tables, or

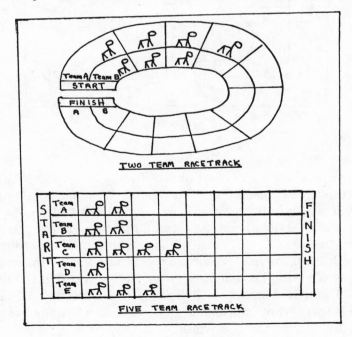

Figure 6-4

sections of the room. In turn, ask a student from each team questions relating to your current math classwork.

Encourage members of the other teams to listen and challenge any answer they believe is incorrect. If the answer is correct, draw a stick figure of a horse in the team's racing lane. Continue playing, making sure all teams have an equal number of turns. The first team to reach the finish line wins.

MATH MACHINES

A simple or complex operation is performed on a number. Students try to guess the operation.

Preparation: None.

Directions:

Draw a diagram representing a machine on the chalkboard; it may be as plain or complicated as you wish.

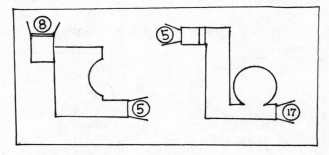

Figure 6-5

Give your class a few examples of how a certain machine works, then ask questions to see if they understand the machine's operation. *Example:* If 8 goes into this machine, 5 comes out. If 5 goes into this machine, 2 comes out. Then ask what happens when 10 goes into the machine. (*Answer:* A 7 comes out; it is a "subtract 3" machine.)

A math machine may perform a combination of operations. Here is an example of a complex machine: When 5 goes in 17 comes out, when 7 goes in 23 comes out, when 3 goes in 11 comes out. What kind of machine is this? What happens if we put 6 into this machine? (*Answer:* This is a 3n + 2 machine; 6 would come out as 20.)

Later have your students draw the machines and make up the rules.

FOOTBALL FOR ALL

A chalkboard version of football in which teams earn yards by answering questions.

Preparation: Questions will need to be prepared and divided into three groups: easy, medium, and hard. Cut a small football shape out of paper and attach tape to it.

Directions:

Draw a diagram of a football field on the chalkboard and tape the football to the 50-yard line.

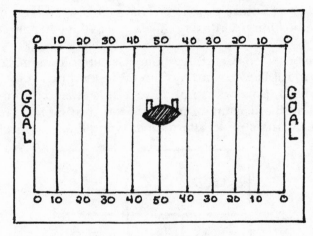

Figure 6-6

Divide the class into two teams and set a time limit. The game begins with a tossup question, which you call out. When a student answers this question, his or her team goes first. After that, teams alternate. Starting at the 50-yard line, teams move the football in opposite directions—one team always moving to the right, the other to the left.

A student may request an easy, a medium, or a hard question (five students may be appointed referees). Easy questions are worth 10 yards, medium questions are worth 20 yards, and the hard questions are worth 30 yards. Once a goal is reached the football is returned to the 50-yard line and the losing team goes first.

Please note that once the tape gets full of chalk, it will no longer stick. Try using two pieces of tape attached to the ends of the football so they will not touch the chalk lines. If necessary, change the tape occasionally.

MATH-TAC-TOE

Students solve nine problems placed in a tic-tac-toe diagram, and earn X's and O's for their team.

Preparation: None.

Directions:

Draw a large tic-tac-toe figure on the chalkboard. Divide the class into two teams—an X team and an O team. Students from these teams alternate coming up to the chalkboard. First, nine students will place a math problem (one at a time) in any of the nine spaces they choose.

Next, other students, in turn, will come up to the chalkboard and answer any one of these problems. Leave a work area on the side of the board available for students who need more space to solve their chosen problem.

If the work is correct (the whole class should agree), that problem is erased and replaced by an X or an O, depending on which team the student is on. If the work is not correct, it is erased and replaced with the original problem again. As in tic-tac-toe, the first team to get three in a row wins. Teams should alternate in going first.

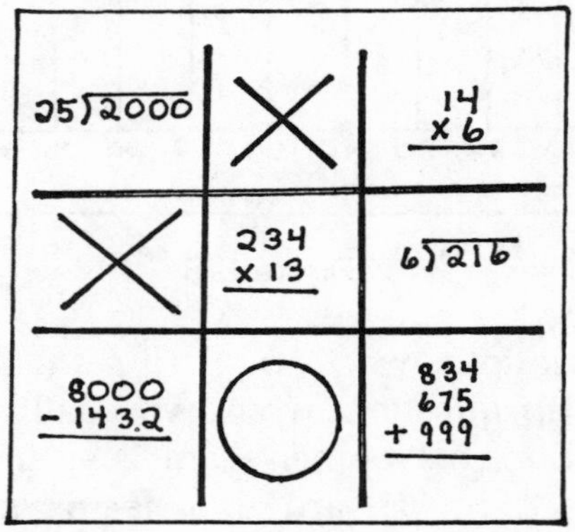

Figure 6-7

MATH STORY

Interesting verbal problems are made up to go with equations.

Preparation: None.

Directions:

Write an equation on the board. For example, $10 - 3 = 7$. Now tell a few stories to illustrate that equation. One could be a simple story of going shopping with 10 dollars, spending 3, and coming home with 7. Another story about the baseball finals might be more complex—how 3 members of the 10-member baseball team were sick on the day of the play-offs and the team lost because they had only 7 players.

Now write another equation on the chalkboard and encourage students to make up stories relating to that equation. At first your students might need a little help in understanding situations where addition, subtraction, or multiplication are used. If this is so, spend some time going over examples before you continue.

Try to get four different stories to illustrate the equation you have written on the chalkboard, then have the class vote on the most interesting story. Try another equation; soon your class should begin to get more comfortable with math story problems.

QUESTIONS THREE

Students answer each other's questions and win cards. A second round may then be played for those who did not win.

Preparation: Hand out blank index cards or small pieces of paper to each student.

Directions:

Have each student write a simple math question on the card or paper. Have students write their names on top of the cards, then collect the cards.

Mix up the cards and begin reading the questions. After a question is read, the first student to give the correct answer is given the card. Students should not answer their own questions.

Continue reading questions until one student has accumulated three cards. This student will come to the front of the room and no longer will answer questions. Instead, he or she may continue reading the next question cards until there is another winner.

After all the cards have been read, congratulate all the three-card winners, two-card winners, and one-card winners. Cards should be collected, and may be used again for another game; this time only students who did not win cards in the first game will participate.

EQUATIONS

Using only the numbers given, students write as many equations as possible.

Preparation: A little planning may be necessary to find numbers that are compatible in forming equations.

Directions:

Write down a series of numbers, for example 6, 8, 4, 2, 12. Students, working independently, write down as many equations as they can within a given time limit using these numbers and only these numbers.

In the above example these equations are possible: $4 + 2 = 6$, $6 - 2 = 4$, $6 + 8 = 12 + 2$, $8 - 6 = 2$, $6 \times 2 = 12$, $8 \times 2 = 12 + 4$, and so on. Decide with your class whether using both $4 + 2 = 6$ and $2 + 4 = 6$ is acceptable.

A point system should be developed awarding one point for equations using three numbers, two points for equations using four numbers, and so on. All equations are checked, and students having the most points win.

REAL-LIFE MATH

Real-life problems using mathematics are discussed in class.

Preparation: None.

Directions:

Share with your class some real-life situations involving math. Depending on the problem, either do all the mathematics yourself or leave some to be done by your class.

For example, at the supermarket you saw a package of four candy bars selling for $1.79; also available was a box containing one dozen bars; costing $4.79. Your students can tell you that this candy sells for 50 cents at the local store. Round out the supermarket prices, and compare the costs.

Pick situations that meet the interest and ability level of your students. As you know, it is important for students to see how math is used in the real world.

In the next activity, you will ask your students to share their math experiences with you.

MATH DIARIES

Your students will write down various experiences they have had using mathematics over a period of a month. This activity is suitable for students in the upper grades.

Preparation: None.

Directions:

Have your students set aside a few pages at the back of their math notebook for their "Math Diaries." As in a real diary, students will write their experiences, in this case, using mathematics. They might discuss their allowances, going shopping with their parents, sharing things with friends, measuring, timing, counting, or comparing.

Positive experiences will probably be written about, such as saving up for something special; but encourage students to share other types of experiences, like a student who went to the store, but didn't know if he had enough money to pay for all the things he wanted.

Give your students an idea of how many items you want (for example, two per week). Discuss the diaries weekly, and at the end of a month decide if this activity should be continued.

CHAPTER 7

How to Use Historic Learning Devices and Methods to Teach and Motivate Today's Mathematics

It has been said many times that we can learn from history. Although this thought was not intended to mean the history of mathematics, it is true that we can teach and motivate our students by looking at some ideas from math history.

For this chapter, topics were selected that deal with place value, addition, multiplication, and geometry in an elementary way. In addition, a look at ancient fractions from three cultures is provided for those students who have already been introduced to formal fractions.

If we wish our students to understand and appreciate our place value system, a look into "Egyptian Numerals" may prove very exciting. Next, a few ancient methods of multiplication are shown: one using fingers, another using addition, and finally one that begins to resemble our own.

Geometry has been fascinating mathematicians for thousands of years. A look back at two Greek mathematicians shows the beginnings of this important subject. Students can also investigate some simple math patterns and systems.

As the teacher, you will need to select the materials that fit into your mathematics programs and the ability level of your students. Illustrations of three abacuses have been included here for such a purpose. The first, the *counting board*, can be used for students just beginning to do basic addition and subtraction. The next, the *simple abacus*, can provide an understanding of place value and regrouping. And finally, the most interesting—the *Chinese abacus*—can offer students further insights and challenges.

We end our historical visit with the beginning of a new branch of mathematics, topology. Perhaps some students will be motivated to delve further into this interesting area.

After each topic in this chapter, problems are suggested for your budding mathematicians to investigate. In many instances a variety of levels are available, designed to enable students with differing abilities and skills to participate in learning about the history of mathematics.

THE ABACUS

One of the earliest mechanical computing devices is the abacus. Originally it consisted of grooves in the sand with pebbles; later it was a wooden frame with beads. The abacus can provide students with an opportunity to understand place value and enable them to gain valuable insights into regrouping.

All three of the abacuses depicted in Figure 7-1 can be inexpensively made with cardboard, a hole puncher, string, and O-shaped cereal or short macaroni, such as ditalini. A rubber band will also be needed for the Chinese abacus.

The counting board has two or more horizontal rows of string with 10 beads (cereal or macaroni) in each row. It can be used for simple addition and subtraction, such as 3 + 4, 9 + 5, or 12 − 7. The beads are used as counters, moving from left to right.

The simple abacus uses three vertical columns, each with 10 beads. The abacus is placed flat on a table. The three columns represent hundreds, tens, and ones. To represent a number such as 46, four beads are pushed up in the tens column and six beads are pushed up in the ones column. Regrouping will be illustrated when 10 beads in the units column are exchanged for one bead in the tens column, as in 46 + 38.

The Chinese abacus has four or more vertical columns, each representing a place value. To make this abacus, draw four vertical lines on a piece of cardboard, as shown in Figure 7-1. Next, punch holes near the top and bottom of each line and with a string thread seven beads (cereal or macaroni) over each line, and tie the string in the back. Finally, place a rubber band horizontally over the cardboard to form a crossbar

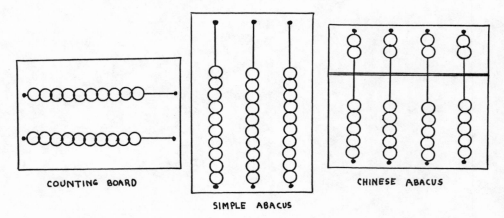

COUNTING BOARD

SIMPLE ABACUS

CHINESE ABACUS

Figure 7-1

separating each column so that two beads are above and five beads are below.

Before using the abacus, place it flat on a table and move all the upper beads up and the lower beads down. The Chinese abacus in Figure 7-1 shows this starting position. The four columns from left to right are thousands, hundreds, tens, and ones.

The two beads above the crossbar each have a value of five, while the beads below the crossbar each have a value of one. So if we wish to represent the number 46, four one-beads are moved up to the crossbar in the tens column; then in the ones column, one five-bead is moved down and one one-bead is moved up (5 + 1 = 6). See Figure 7-2(a).

Now suppose we wish to add 38 to 46. Let us begin by adding 8 to the ones column. We would move one more five-bead down, and three one-beads up (5 + 3 = 8), as in Figure 7-2(b). In the ones column, we now have two five-beads and four one-beads at the crossbar.

At this point we can exchange two fives for a 10. We do this by moving the two five-beads in the ones column back up to the top, and then moving a one-bead in the tens column up towards the crossbar. See Figure 7-2(c).

Finally we need to add three tens. This could usually be done by moving three one-beads up in the tens column. However, all the one-beads in the tens column are already up, so we can use this little trick: Since 3 = 5 − 2, we can add a five-bead and subtract two one-beads. The answer, 84, can be seen in Figure 7-2(d).

Here is another shortcut your students might be interested in. If we wished to add nine tens (90) to the previous sum of 84, knowing that 90 = 100 − 10, we could move a one-bead up to the crossbar in the hundreds column, and then move a one-bead back down in the tens column.

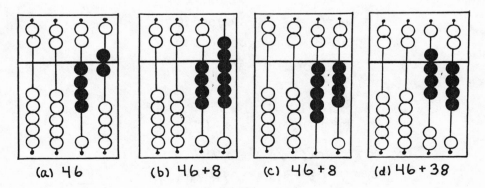

(a) 46   (b) 46+8   (c) 46+8   (d) 46+38

Figure 7-2

It would probably be best for your students to start on easier problems before proceeding to the example outlined in Figure 7-2, such as 26 + 13, 254 + 135, and 126 + 257.

The Chinese abacus is fascinating to use. As students become adept at it, they can take advantage of the many shortcuts available for speeding up calculations. Once addition is understood you might challenge your students to try subtraction examples on the abacus.

A historical note: the word abacus comes from the Greek word *abax,* meaning sand tray; and the words calculate and calculus come from *calculi,* the Latin word for pebble—thus a reminder of the original abacus, a series of grooves in the sand with pebbles.

EGYPTIAN NUMERALS

The colorful but cumbersome Egyptian hieroglyphic number system, over five thousand years old, can help our students to appreciate the efficiency of our present place value system.

The Egyptian number system uses a different symbol for each power of ten; each symbol being repeated as often as necessary. The symbol of 1 is a vertical staff, 10 is a heel bone, 100 is a scroll, 1000 a lotus flower, 10,000 a pointing finger, 100,000 a fish, and 1,000,000 is a man with hands up in astonishment. Figure 7-3 shows how hieroglyphic notation is used to represent several large numbers.

The last Egyptian numeral shown, 1986, uses 24 symbols to represent a number we represent with only four symbols. Certainly Egyptian numeration is not more efficient than our number system, but perhaps your students might find it more fun.

What would today's date look like in Egyptian hieroglyphics?

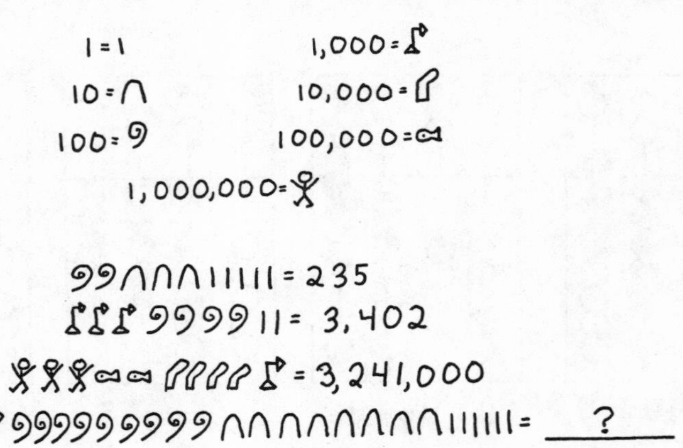

$$1 = 1 \qquad\qquad 1,000 = ℓ$$
$$10 = \cap \qquad\qquad 10,000 = ℓ$$
$$100 = 9 \qquad\qquad 100,000 = ⌒$$
$$1,000,000 = ⚡$$

$$99\cap\cap\cap|||||= 235$$
$$ℓℓℓ\,9999\,|| = 3,402$$
$$⚡⚡⚡⌒⌒\;ℓℓℓℓ\;ℓ = 3,241,000$$
$$ℓ\,999999999\,\cap\cap\,\cap\cap\cap\cap\cap|||||| = \underline{\quad?\quad}$$

Figure 7-3

FINGER MULTIPLICATION

This was an old method used by European peasants. By using our fingers, we can find the multiplication facts for the numbers from 6 to 10. It is interesting to do, but best done by students who are already familiar with their multiplication facts. Students will enjoy seeing the products they expected appear on their fingers.

Finger multiplication takes three simple steps. First students hold their hands up as shown in Figure 7-4. Notice that each finger represents a number from 6 to 10.

Now if we wish to multiply 8 × 7, we would let finger 8 touch finger 7, and bend all the fingers below these touching fingers forward.

Finally, we need only to read the results. The tens digit of the product is found by counting the number of extended fingers. Each of these fingers is worth 10. The units digit is found by multiplying the number of bent fingers on the left hand by the number of bent fingers on the right hand.

hand position

8 × 7

Figure 7-4

In our example there are five extended fingers, so the tens digit is 5. There are two bent fingers on the left hand and three bent fingers on the right hand. We multiply 2 times 3 to get 6, the units digit. Our product is 50 + 6 = 56. It is just as we expected, 8 × 7 = 56.

In a problem such as 6 × 7, we will have three extended fingers for a value of 30. Our bent fingers will give us 4 × 3 = 12 for the units value. Combining these, the product becomes 30 + 12 = 42.

Another unusual result is found in 10 × 6. Our fingers will give us 60 + (0 × 4) = 60.

Have your students try 7 × 7, 6 × 8, 9 × 6, 8 × 8, 9 × 8, and 10 × 10. This time it will be all right for students to do math on their fingers.

MULTIPLICATION BY DOUBLING

As we have seen, Egyptian numbers tended to become lengthy at times. It would have been difficult for them to develop a system of multiplication similar to ours. Instead they developed a method capable of multiplying any size numbers, using only two simple operations—doubling and adding.

We will multiply 26 × 214 using the Egyptian method of doubling. First we write the number 1 in one column and the number 214 in another column. We will keep doubling the numbers in both columns until there are enough numbers in the first column so that we might get a sum of 26:

	1	214	
*	2	428	428
	4	856	
*	8	1712	1712
*	16	3424	3424
	26		5564

The rows marked with an asterisk are those with numbers having a sum of 26. We may cross out the other rows and only add the numbers in the asterisk-marked rows. Since 2 + 8 + 16 = 26, we add 2 × 214 = 428, 8 × 214 = 1712 and 16 × 214 = 3424 to find our product: 26 × 214 = 5564.

Here is a final example illustrating multiplication by doubling: 41 × 62.

	1	62	
*	1	62	62
	2	124	
	4	248	
*	8	496	496
	16	992	
*	32	1984	1984
	41		2542

Some examples your students might try are 11 × 15, 12 × 12, 20 × 64, and 27 × 121.

LATTICE METHOD OF MULTIPLICATION

This method of multiplication more closely resembles our own. By placing numbers in squares with diagonals and then adding diagonally, regrouping (carrying) may be avoided until the final step.

In the lattice diagram, one factor is written horizontally and the other factor is written vertically. The number of rows and columns in the lattice will depend upon the number of digits in the factors.

Once the diagram is drawn, we will multiply each of the digits and place the products in the appropriate squares, as shown in Figure 7-5. Note that in each square the diagonal separates the tens from the units digit.

After all the products are placed within the lattice, we add all the digits along the diagonal columns. Two examples are given below. The first, $56 \times 12 = 672$, requires no regrouping, while $425 \times 38 = 16{,}150$ uses carrying only in the final step.

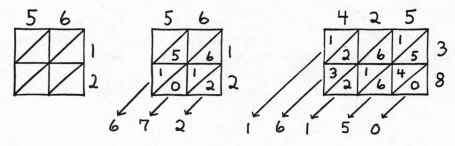

Figure 7-5

ANCIENT FRACTIONS

In our fractions, any number may be the numerator and any number (except zero) may be the denominator. Yet historically this was not always so. Your students might better appreciate our system of fractions, which took 3000 years to develop, after a look at some ancient fractions.

Egyptian fractions: The Egyptians only used fractions that had a numerator of 1. Any denominator was allowed. A fraction with 1 as the numerator is called a unit fraction. To show a unit fraction they would place a symbol (an ellipse) over the hieroglyphic number they wished to use for the denominator. Therefore an ellipse over three vertical lines meant ⅓, and an ellipse over a heel bone represented ⅒.

If a fraction was needed that did not have one as a numerator, it was then written as the sum of unit fractions. For example: ¾ = ½ + ¼; ²⁄₇ = ¼ + ¹⁄₂₈.

To see how your class might break down a fraction into unit fractions, show them this example: ⁷⁄₂₄ = ³⁄₂₄ + ⁴⁄₂₄ = ⅛ + ⅙. Although not all fractions are this easily converted, let your students try this method on $\frac{7}{12}$, $\frac{8}{15}$, $\frac{9}{20}$, and $\frac{10}{21}$.

Babylonian fractions: Since the Babylonians used a base 60 numerical system (as compared with our base 10 system), it is not surprising that they based their fractions on 60. All fractions had to have 60 as a denominator, or a multiple of 60 (60 × 60 = 3600). However, any number could be used for the numerator.

As you might suspect, this led to very complicated fractions. Consider that $\frac{3}{8} = \frac{22}{60} + \frac{30}{3600}$. Now your students might think that they would not want any part of Babylonian fractions, but actually we use them whenever we tell time. There are 60 seconds in a minute, 60 minutes in an hour, and also 3600 seconds in an hour.

Roman fractions: All Roman fractions had 12 or multiples of 12 as their denominators. This was based on the Roman weight and money systems. Here is what Roman fractions looked like:

$(\frac{1}{12})$ – $(\frac{5}{12})$ = = – $(\frac{9}{12})$S = –

$(\frac{2}{12})$ = $(\frac{6}{12})$ S $(\frac{10}{12})$ S = =

$(\frac{3}{12})$ = – $(\frac{7}{12})$ S – $(\frac{11}{12})$ S = = –

$(\frac{4}{12})$ = = $(\frac{8}{12})$ S = $(\frac{12}{12}$ equals 1) I

You can ask your students how the Romans would have written: $\frac{1}{2}$, $\frac{1}{3}$, $\frac{1}{6}$, $\frac{3}{4}$, $\frac{2}{3}$, and $\frac{5}{6}$. What familiar measurements are divided into 12 units? (12 inches in a foot, 12 units in a dozen, 12 months in a year.)

SIEVE OF ERATOSTHENES

A prime number is a number that has no factors other than itself and one. Three and five are primes, four is not since 4 = 2 × 2. Which numbers are primes, which are not? Over 2000 years ago a Greek mathematician named Eratosthenes developed a simple method for finding prime numbers.

1. Write down a sequence of integers, beginning at 2. (1 is not considered a prime.)

2   3   4   5   6   7   8   9   10   11   12   ...

2. Circle the first number 2 and cross out all multiples of 2.

②   3   4̸   5   6̸   7   8̸   9   1̸0̸   11   1̸2̸   ...

3. Circle the next unmarked number and cross out all multiples of that number which have not yet been crossed out.

②   ③   4̸   5   6̸   7   8̸   9̸   1̸0̸   11   1̸2̸   ...

4. Go to the next unmarked number, circle it and cross out its multiples. Continue until all numbers have been marked. All the circled numbers are the primes.

②   ③   4̸   ⑤   6̸   ⑦   8̸   9̸   1̸0̸   ⑪   1̸2̸   ...

You might wish your class to investigate all prime numbers up to 100 (or 50) by using the sieve of Eratosthenes. But first let them guess how many primes there are, and which are the prime numbers.

SQUARE AND TRIANGULAR NUMBERS (PYTHAGORAS)

To the Greek mathematician Pythagoras, numbers were very important. He believed that the secrets of nature could be discovered by studying whole numbers, and he founded a secret society that met to discuss these ideas.

Pythagoras liked to think of numbers in terms of their shapes. He could study numbers by putting pebbles together into various geometric shapes. The two most important shapes to the Pythagoreans were triangles and squares. Triangle numbers looked like this:

1   3   6   10

Pythagoras noted that triangular numbers were the sums of consecutive integers: $1 = 1, 3 = 1 + 2, 6 = 1 + 2, + 3, 10 = 1 + 2 + 3 + 4$. Ask your class to find the next triangular number. Is it the sum of consecutive integers?

Square numbers were the sums of consecutive odd numbers. For example, 16 = 1 + 3 + 5 + 7. Here are the first four square numbers:

    .   . .   .
    1   . .   . .
               4   . . .   .
                           9   . . . .
                                       16

You might ask your students to find the next square number. Is it the sum of odd integers? Note that there is a relationship between the square and triangular numbers shown here, 4 = 1 + 3, 9 = 3 + 6, 16 = 6 + 10. Each square number seems to be the sum of two triangular numbers. Is this always true? Let your students investigate.

SIMILAR TRIANGLES (THALES)

Thales, the first known Greek mathematician, measured distances by using similar triangles. The story is told that Thales traveled to Egypt and amazed the Egyptian king by determining the height of the Great Pyramid by measuring its shadow. Your students may be able to use this method to find the height of tall objects such as flagpoles and trees.

Similar triangles are triangles having exactly the same shape, but not necessarily having the same size. The corresponding angles are equal, and the corresponding sides are in proportion. There are two ways to use similar triangles to find the height of an object.

In one version of the story, Thales set a stick vertically in the ground and watched its shadow. When Thales noted that the length of

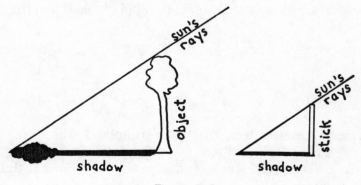

Figure 7-6

the stick's shadow was equal to the height of the stick, he measured the length of the pyramid's shadow. He knew that the length of the shadow must be equal to the height of the pyramid.

In another story, Thales did not wait for the shadow to be equal to the stick, but found the height by using proportions—since the sides of similar triangles are in proportion:

$$\frac{\text{Height of Object}}{\text{Shadow of Object}} = \frac{\text{Height of Stick}}{\text{Shadow of Stick}}$$

Perhaps it will be possible for your class to find the height of some tall object by using one of these methods. Note that the length of a shadow will vary according to the time of day and the month of the year, with shorter shadows in summer and longer ones in winter.

ADDING 1 TO 100 (GAUSS)

Your class might appreciate this story about Karl Frederick Gauss. When Karl was a young boy in school, his teacher gave the class an assignment that the teacher supposed would keep his students busy for a long time: to add up all the numbers from 1 to 100. To the teacher's astonishment, young Karl had the answer in a few seconds—5050. How did Gauss arrive at this answer?

While Gauss used multiplication in his solution, a few simple examples will be given later that can be done entirely with addition. Let us look at Gauss's solution. Note the sum of the first and last number, the second and next to last number, etc.

$$1 \ + \ 2 \ + \ 3 \ + \ 4 \ + \dots + \ 97 \ + \ 98 \ + \ 99 \ + \ 100$$

$$101$$
$$101$$
$$101$$

Each number pair adds up to 101: 1 plus 100, 2 plus 99, 3 plus 98, 4 plus 97, all the way up to 50 plus 51. Since there are 100 numbers, there are 50 number pairs, each with sum of 101. Thus the sum of the numbers from 1 to 100 would be $50 \times 101 = 5050$.

Here are some examples that might be done using only addition, such as adding the even numbers from 2 to 12.

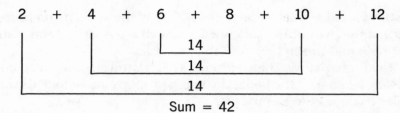

$$2 \; + \; 4 \; + \; 6 \; + \; 8 \; + \; 10 \; + \; 12$$

Sum = 42

Have your class try these simple problems. Perhaps they can use both regular arithmetic and Gauss's method and compare results.

1) Find the sum of 2 + 4 + 6 + 8.

2) Add the first 6 odd numbers: 1, 3, 5, 7, 9, 11.

3) Add all the integers from 1 to 10.

Here are some problems that are best done with Gauss's method. Students will need to use both addition and multiplication, and be aware of how many number pairs are involved.

1) Add all the integers from 1 to 20.

2) What is the sum of the numbers from 1 to 60?

3) Find the sum of the even numbers between 2 and 100. (Remember there are 50 even numbers here, so there are 25 number pairs.)

For students who did not have any difficulty with the above examples, pose the following question: So far we have worked on problems that always contained an even number of integers. How would you use this method on problems containing an odd number of integers? a) add 2 + 4 + 6 + 8 + 10, or b) add the integers from 1 to 25.

PASCAL'S TRIANGLES

This well-known arithmetic triangle was constructed by Pascal over 300 years ago, using only simple addition. Yet there are many interesting relationships here, involving addition, multiplication, algebra, and probability. Your students might enjoy making Pascal's triangle, and looking for addition or multiplication relationships.

```
              1
           1     1
        1     2     1
     1     3     3     1
   1    4     6     4     1
 1    5    10    10     5     1
 —    —    —     —     —     —     —
```

Above are the first six rows of Pascal's triangle. Each number in the triangle is found by adding the two numbers to the left and right in the line above. If we wished to find the seventh row, the first number would be 1 (since the only number above it is 1), the second number would be 6 (1 plus 5), the third number would be 15 (5 plus 10), the fourth would be 20 and so on.

For addition relationships:

1) Give your students the first five rows of Pascal's triangle, and ask them to find the sixth and seventh rows.

2) Have your students find the sum of all the numbers in each row. For example, the sum of the numbers in row six is 32. What relationship can they find about these sums? (The sum of each row is exactly twice the sum of the preceding row.)

For multiplication relationships, let your class expand Pascal's triangle to the first nine rows.

1) Circle all even numbers (multiples of two) and look for a pattern.

2) This time circle only multiples of three; a different pattern should be formed.

3) Try again, using only multiples of five.

There are many other relationships in Pascal's triangle. If your class investigated triangular numbers mentioned earlier in this chapter, they can find all of these numbers along one of the diagonal lines. Square numbers can be found by adding two adjacent numbers along this same diagonal: $1 + 3 = 4$, $3 + 6 = 9$, $6 + 10 = 16$, and so on.

The diagonal before it represents all the counting numbers and the diagonal after it represents three-dimensional triangles—triangular pyramids.

Can your students find any other relationships? Let your students add the first two, three, or four numbers along any of the diagonals. Are these sums always found in Pascal's triangle?

BRIDGES OF KÖNIGSBERG (TOPOLOGY)

In the 1700s the people of the city of Königsberg used to go for walks around their city. It became a challenge to find a path that would cross all seven bridges without having to go over any bridge twice.

But no matter how often they tried, no one in Königsberg could find such a path. Perhaps your class might like to try, using the map in Figure 7-7.

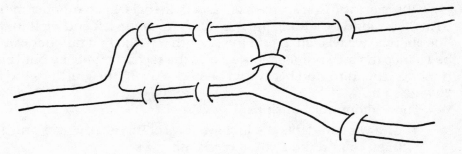

Figure 7-7

When the Swiss mathematician Euler heard of this problem, he decided to investigate. He found that it was impossible to plan a walk so that each bridge was crossed only once. His study of this problem led to the beginning of a new area of geometry, later to be called topology.

It was not until much later, when an eighth bridge was built, that it was possible to walk across each bridge once. A map of the river and its eight bridges is shown in Figure 7-8. See if your class can take this famous topological walk on the bridges of Königsberg. *Hint:* If you start at one of the areas marked with an "X" and end at the other "X," it will be possible to cross each of the bridges only once.

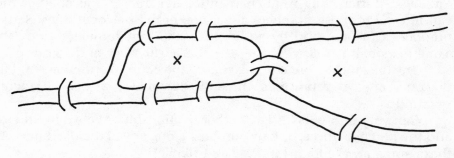

Figure 7-8

8

Puzzles, Tricks, and Patterns: Explorations With or Without the Electronic Calculator

Our explorations start with one of the oldest known puzzles, the magic square, and end with investigations of number patterns—best done with our present-day electronic calculators. In between are problems old and new, some that can be used for entertainment, some for enrichment, and some for the challenge.

Most teachers feel that calculators have a place in the classroom, but that students must first learn basic skills—addition, subtraction, multiplication, division. Calculators can then be used to check problems, to search for number patterns, and to develop and investigate ideas.

All the puzzles, tricks, and patterns in this chapter can be done with paper and pencil, if practice in basic skills is deemed necessary. However, calculators would be appropriate in the longer problems where students are searching for ideas or patterns.

There is something here for students of all ability levels, from the simple addition of "Magic Triangles" to the powers of two in the "Chessboard Problem."

The seven number tricks in this chapter might be supplemented by two geometry activities in the next chapter, "Möbius Strips" and the "Four-Sided Paper." Both of these make excellent "magic tricks."

MAGIC SQUARES

(addition)

The magic square is one of the oldest number puzzles known. It was believed to be a source of magic, since the sum of each row, each column, and each diagonal is the same.

The original problem was to place the digits 1 to 9 in the squares of Figure 8-1(a) in such a way that the sum of each row, column, and diagonal equals 15. One possible solution is shown in Figure 8-1(b).

You might wish to put in a few of the numerals for your students and allow them to fill in the remaining squares. This method was used by teachers in India over a thousand years ago.

Magic squares have been made larger and inside numbers have been changed. Benjamin Franklin created a special 8 by 8 magic square. While this size would probably not be appropriate for most students, there is a way to vary magic squares.

Using the familiar 3 by 3 square in Figure 8-1(b), each number is then changed by adding a constant—for example 5. The square in Figure 8-1(c) uses the numbers 6 to 14, and each row, column, and diagonal has a sum of 30. Can your students complete this magic square?

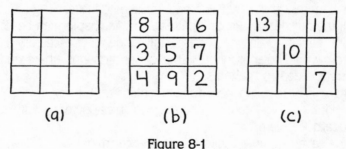

(a)   (b)   (c)

Figure 8-1

MAGIC TRIANGLES

(addition)

In magic triangles, the sums of the numbers on each side are equal. Different arrangements of the numbers will result in different sums:

1) The numbers 1 to 6 are to be placed in the six circles of the triangle shown in Figure 8-2(a), so that the sum of each side is 10. (Hint: the three corner numbers are 1, 3, and 5.)

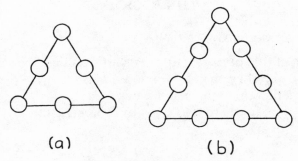

Figure 8-2

2) Next, the numbers 1 to 6 can be placed so that the sum of each side is 9. (Hint: the corners are 1, 2, and 3.)

3) This magic triangle can be arranged so that the sums equal 12. (Hint: the corners are 4, 5, and 6.)

Using the larger triangle, Figure 8-2(b), the numbers 1 to 9 can be arranged in the nine circles so that the sum of each side is 20. (Hint: the three corners are 4, 5, and 6.) These nine digits can also be arranged to give sums of 17 (corners—1, 2, 3) or sums equal to 23 (corners—7, 8, 9).

MAGIC STARS

(addition)

Can your students place the numbers 1 to 5 in the circles of the star below, Figure 8-3(a), so that the sum of the four numbers along each line is 12?

In the star in Figure 8-3(b), the four numbers on each of the five lines should have a sum of 30. Again, use the numbers 1 to 5.

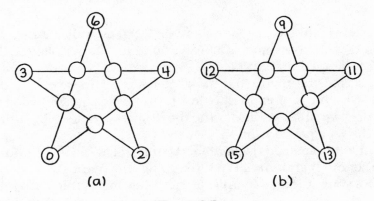

Figure 8-3

THE FOUR 4's

(all operations, fractions)

The idea of this puzzle is to represent numbers by using one, two, three, or four 4's. They may be combined using $+$, $-$, $\times$, $\div$, or any combination. Fractions may also be used.

For example, here are two ways to represent 12:

$$12 = 4 \times 4 - 4 \qquad \text{and} \qquad 12 = \frac{44 + 4}{4}$$

Can your students find ways to represent the numbers from 1 to 12? [$1 = 4 \div 4$, $2 = (4 + 4) \div 4$, $3 = 4 - (4 \div 4)$, and so on.]

CHECKERBOARD SQUARES

(geometry)

An ordinary checkerboard has eight rows of squares with eight squares in each column. Ask your students how many squares there are in a checkerboard. Let them consider not only the 64 1 by 1 squares, but the larger 2 by 2 squares, 3 by 3 squares, and up to the large 8 by 8 square. Some of these may overlap.

This is not an easy problem, and may have to be done on the chalkboard where overlapping squares can be kept track of.

The answer is 204, made up of one 8 by 8, four 7 by 7's, nine 6 by 6's, 16 5 by 5's, 25 4 by 4's, 36 3 by 3's, 49 2 by 2's, and 64 1 by 1's.

THE CHESSBOARD PROBLEM

(doubling)

There is an old legend about a King of India. When he was presented with the game of chess, the king was so pleased that he wanted to reward the inventor.

He asked the man what he wanted for his reward. The inventor replied "Give me one grain of wheat for the first square, two grains for the second square, 4 grains for the third, and keep doubling until every one of the 64 squares is filled."

While this seemed like a modest proposal, it turned out that there was not enough grain in all of India to fill this request.

Have your students try this project. They might think of a penny for the first square, two pennies for the next, then four pennies, eight pennies, 16 pennies, and so on. Let them estimate how much money they think it would take to fill the chessboard.

Soon after starting, your students should notice how large the numbers are becoming. At some point you might wish to stop your students and write this number on the board: 18,446,744,073,709,551,615. That is the number of pennies it would take to fill all 64 squares—over 18 quintillion.

MAZES

(all operations)

Figure 8-4 illustrates two mazes for different ability levels. Your students will need paper and pencil. In either maze, the students begin in the START square and move to adjacent squares—right, left, up, and down, but not diagonally.

They do the mathematics along the way, and the object is to reach the END square with the correct results.

The correct path in Maze A is $5 \times 3 + 6 \div 7 - 1 \times 8 + 4 = 20$. There are two correct paths that equal 500 in Maze B.

Next, you might encourage your students to make their own mazes to be shared with the class.

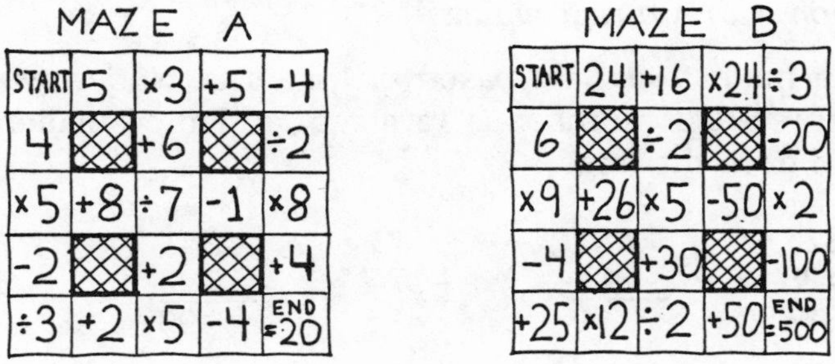

Figure 8-4

BINARY CARD TRICK

(addition)

After your students prepare the four cards shown in Figure 8-5, they can perform the following trick. (Note that the first numbers on the cards are 1, 2, 4, and 8, respectively. These are numbers of the binary system.)

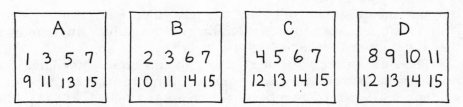

Figure 8-5

A student places the cards on a table and asks a friend to think of a number between 1 and 15.

The friend is then asked to point to the cards which contain this number. Your students can find the number simply by adding the first number on each of the cards pointed to.

For example, if the friend said the number was on cards A, B, and D, student adds 1 + 2 + 8, and gets the sum of 11.

This trick is based on the binary system. Each number is placed on the cards according to its binary representation. For example, 5 is written as 101 in the binary system, so it is placed on the first and third cards.

SUBTRACTION TRICK

(subtraction with two regroupings)

Some math tricks are based on patterns. Notice what happens when a three-digit number is reversed, and the smaller number is subtracted from the larger:

$$
\begin{array}{ccccc}
824 & 483 & 671 & 251 & 956 \\
-428 & -384 & -176 & -152 & -659 \\
\hline
396 & 99 & 495 & 99 & 297
\end{array}
$$

Either (a) the answer is 99, or (b) the answer is a three-digit number where 9 is the middle digit and the first and third digits have a sum of 9.

This knowledge can be used as follows. Ask a student to come to the chalkboard and write any three-digit number in which the first digit is larger than the last. You should not look at the chalkboard during the trick.

Now have the student reverse the digits and subtract. The class should check that the subtraction was correctly done.

Ask the student to tell you the first digit of the answer, and from that you will announce the entire answer. If the first digit is 9, that is case (a) and the answer must be 99. If the first digit is any other number, the answer must be a three-digit number as in case (b). Thus, if the first digit is 2, the middle digit must be 9, and the last digit must be 7. Remember that the first digit and the last digit must always have a sum of 9.

Repeat with different three-digit numbers until your students discover how this trick is done.

FAVORITE NUMBER TRICK

(all operations)

If your students can add, subtract, multiply, and divide (division only by 2 is needed), they can try this trick on their friends. To illustrate this favorite number trick, ask some students to:

1) Write down their favorite number.   (x)

2) Multiply this number by 4.   (4x)

3) Add 20 to the product.   (4x + 20)

4) Divide this final sum by 2.   (2x + 10)

Call on students one at a time. Ask them to tell you only the final result after the division.

From each student's results, you will be able to find their favorite number by these simple steps:

a) Take the final result.   (2x + 10)

b) Divide it by 2.   (x + 5)

c) Then subtract 5.   (x)

This is, of course, the original number. *Example:* 1) favorite number is 12, 2) 12 × 4 = 48, 3) 48 + 20 = 68, 4) 68 ÷ 2 + 34; a) final result is 34, b) 34 ÷ 2 = 17, c) 17 − 5 = 12.

BIRTHDAY TRICK

(addition, subtraction, multiplication)

In this trick your students can guess a friend's birthday. Your student would tell a friend to follow steps one through eight:

1) Write the number of the month that your
   birthday is in. For example: April is 4, July
   is 7.   (Let m = month)

2) Multiply that number by 5.   (5m)

3) Next add 6.   (5m + 6)

4) Multiply by 4.   (20m + 24)

5) Then add 3.   (20m + 27)

6) Multiply by 5.   (100m + 135)

7) Add the day of the month on which you were born. For example, if birthday is July 28, add 28.   (let d = day)

8) Give me your final result.   (100m + 135 + d)

Now your student subtracts 135 from the results and gets the friend's birthday (100m + d). The first digit or two tells the month, the last two digits represent the day. For example, 728 is July 28, 1203 is December 3, and 1023 is October 23.

MULTIPLICATION TRICK

(multiplication of an eight-digit number)

To perform this trick it is necessary to accurately multiply an eight-digit number by a two-digit number.

First select a volunteer who then tells you his or her favorite number between 1 and 9. To yourself, multiply this favorite number by 9. Take the product and ask to have 12,345,679 multiplied by that product.

For example, if you were told that the favorite number is 7, you would ask to have 12,345,679 x 63. Watch what happens if 7 is the chosen number:

$$
\begin{array}{r}
12345679 \\
\times 63 \\
\hline
37037037 \\
74074074 \\
\hline
777777777
\end{array}
$$

This trick will work as well for any multiple of 9. The key to the trick is that $12{,}345{,}679 \times 9 = 111{,}111{,}111$. Therefore, using any multiple of 9 will result in that number times 111,111,111 as the final product.

DIVISION TRICK

(division by a two-digit number)

Have a volunteer pick a three-digit number and tell it to you. Then have it written twice to form a six-digit number. For example, 524 would become 524,524.

Announce that you will ask to have three divisions done, and that you will not look at any of the calculations, but that you are now writing down the answer that will be the final result. This claim might be met with disbelief, but the trick is really very simple. On a piece of paper you simply write down the original three-digit number. Fold the paper and ask that it not be opened until the three divisions have been completed.

First ask to have the six-digit number divided by 7, then that result divided by 11, and finally that second result divided by 13. Your prediction may now be opened and it will be the correct final result.

Let us work out the divisions for 524: 1) 524524 ÷ 7 = 74932; 2) 74932 ÷ 11 = 6812; and 3) 6812 ÷ 13 = 524.

Note that all divisions should work out exactly, leaving no remainders. Therefore, a remainder at any step would imply that an error has been made.

The reason this trick works so nicely is that 7 × 11 × 13 = 1001, and that any number in the form ABCABC when divided by 1001 will equal ABC.

QUICK TRICK

(multiplication)

Challenge your students to find the product of the following 10 numbers. They may use calculators or paper and pencil.

As soon as they know the correct answer, ask them to write it down and raise their hands, but ask them not to tell anyone their answer.

$$9 \times 8 \times 7 \times 6 \times 5 \times 4 \times 3 \times 2 \times 1 \times 0$$

The answer, of course, is zero; and it was not necessary to do any multiplication at all.

PATTERNS: 10 INVESTIGATIONS

(multiplication, division, fractions)

These investigations can be done with paper and pencil or with the help of an electronic calculator. Can your students find the patterns in each?

(1) 37 × 3 =
    37 × 6 =
    37 × 9 =
    37 × 12 =
    37 × 15 =
    37 × 18 =
    37 × 21 =
    37 × 24 =
    37 × 27 =

(2) 91 × 11 =
    91 × 22 =
    91 × 33 =
    91 × 44 =
    91 × 55 =
    91 × 66 =
    91 × 77 =
    91 × 88 =
    91 × 99 =

(3)   6 x 7 =
       66 x 67 =
      666 x 667 =
     6666 x 6667 =
    66666 x 66667 =

(4) 7 x 15873 =
    14 x 15873 =
    21 x 15873 =
    28 x 15873 =
    35 x 15873 =

(5)   81 ÷ 9 =
             882 ÷ 9 =
           8,883 ÷ 9 =
          88,884 ÷ 9 =
         888,885 ÷ 9 =
       8,888,886 ÷ 9 =
      88,888,887 ÷ 9 =
     888,888,888 ÷ 9 =
   8,888,888,889 ÷ 9 =

(6)   1 ÷ 9 =
               11 ÷ 9 =
              111 ÷ 9 =
            1,111 ÷ 9 =
           11,111 ÷ 9 =
          111,111 ÷ 9 =
        1,111,111 ÷ 9 =
       11,111,111 ÷ 9 =
      111,111,111 ÷ 9 =

(7) $\frac{1}{9}$ =

$\frac{2}{9}$ =

$\frac{3}{9}$ =

$\frac{4}{9}$ =

$\frac{5}{9}$ =

$\frac{6}{9}$ =

(8) $\frac{1}{11}$ =

$\frac{2}{11}$ =

$\frac{3}{11}$ =

$\frac{4}{11}$ =

$\frac{5}{11}$ =

$\frac{6}{11}$ =

(9) $\frac{1}{7}$ =

$\frac{2}{7}$ =

$\frac{3}{7}$ =

$\frac{4}{7}$ =

$\frac{5}{7}$ =

$\frac{6}{7}$ =

(10) 142857 x 1
     142857 x 2
     142857 x 3
     142857 x 4
     142857 x 5
     142857 x 6

continue   continue   see note on next page

Please note that the patterns formed in the last two problems are different from the other more predictable patterns. In these results, the same digits always appear in the same order, but each time the initial digit is different. These are known as cyclic numbers.

9

Geometry, Logic, and Creativity

The three topics in this chapter are useful in establishing a well-rounded mathematics program. After all, the subject is mathematics and we should present it as more than arithmetic facts and operations. In mathematics there are decisions to be made; there is the need for logic and the room for creativity.

Geometry itself was founded and developed by a combination of logic and creativity. Mathematicians from Archimedes to Gauss have told stories of solving difficult geometry problems with sudden creative insights. This was both preceded and followed by the more usual logical approach to mathematical problem solving.

In "Shape, Color, and Size," there are decisions to be made as geometry is combined with logic. Making geometric designs uses both creativity and geometry, while "Which Is Different" combines creativity and logic to create a game adaptable for students of all ability levels.

All the geometry, logic, and creative activities and games in this chapter can be interspersed into your regular math curriculum. The logic games may be repeated throughout the school year as students start to develop his or her own strategies. A few suggestions on creative activities have been offered, but each teacher should expand on these—choosing activities suited to his or her students' individual abilities.

GEOMETRY PUZZLES

(geometry)

Puzzles can be made for the geometry shapes being studied by your students. Cut out four familiar geometric figures, then cut each shape into two pieces. (A suggestion is given in Figure 9-1.) Then mix all eight pieces together.

Students should be given the eight pieces and asked to sort them into a circle, a square, a triangle, and a rectangle (or any other chosen shapes).

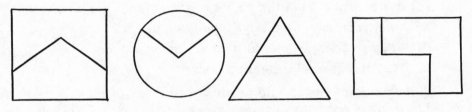

Figure 9-1

JUST-ONE-CUT

(geometry, logic)

For a more challenging geometric puzzle, have your students try the four figures shown in Figure 9-2. Copy only the solid lines, not the dotted ones, and present these figures to your class. (If possible place the figures on one sheet of paper and duplicate for each student.)

Your students should be told that each of the figures can be made into a square by cutting each into two pieces and then rejoining these pieces in a different arrangement.

The dotted lines above represent the cuts necessary for the above figures. If you wish to invent new figures for this activity, start by cutting

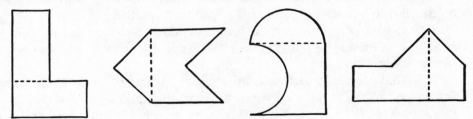

Figure 9-2

out a square. Then cut the square into two pieces, and place these together in a different relationship. Finally, trace the resulting shape.

SHAPE, COLOR, AND SIZE

(geometry, logic)

For each set, cut out shapes of different colors and sizes. A set could contain three different shapes—squares, triangles, and circles; three different colors—red, yellow, and blue; and three different sizes—small, medium, and large.

For example, a set of nine pieces might contain the following:

RED: Large square, medium triangle, small circle

BLUE: Large triangle, medium circle, small square

YELLOW: Large circle, medium square, small triangle.

Activity I: Students should be shown these assorted shapes, then asked to think of a way to sort these pieces into three different groups. After one way is found and the pieces sorted, ask for a completely different way of sorting. Finally, ask if there is yet another way to sort these shapes. (Students should discover groupings by color, by shape, and by size.)

Activity II: Out of the nine assorted shapes, choose two of the shapes that have one characteristic in common. *For example:* large *red* square and small *red* circle or medium blue *circle* and large yellow *circle.* Ask students to find another shape that could belong to this group.

DESIGNING WITH A COMPASS—HEXAGONS AND FLOWERS

(geometry, creativity)

Once your students are able to draw circles with a compass, they might like to draw hexagons and additional creative designs.

Hexagons may also be used for math games mentioned in previous chapters, such as the game called *Beehive* in Chapter 2, and the game *Hexagons* in Chapter 3.

To make a hexagon, first draw a circle with a compass. Next, using the same radius, make a series of arcs around the circumference. (See Figure 9-3.) It is important that the same opening of the compass (the radius) used to draw the circle is used throughout the construction.

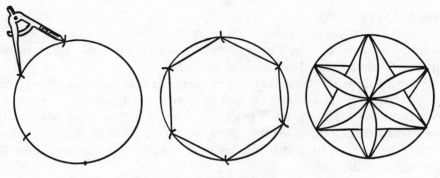

Figure 9-3

Six arcs should fill the circumference evenly, and if we connect these we will have a perfect hexagon.

By extending these arcs further into circles, then adding lines, other circles and color—numerous designs may be created.

POLYOMINOES

(geometry, logic)

Polyominoes can be used as a game, a puzzle, a lesson in perimeter and area, or as an activity with three-dimensional shapes.

Your students are all familiar with the shape of a domino—two squares side by side. If we wish to use three squares (trominoes), how could we arrange them?

A1, A2, B1, B2, B3, and B4 in Figure 9-4 show many arrangements for three squares. However, of all these arrangements there are only two

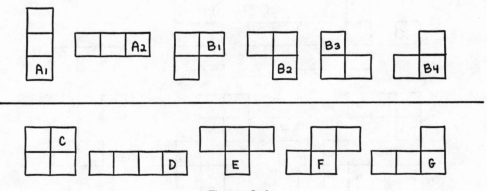

Figure 9-4

different shapes. A1 and A2 represent the same shape; B1, B2, B3, and B4 are all rotations and reflections of each other. Have your students try to find the five different arrangements of four squares (tetrominoes). Figures C to G show these arrangements.

Now that your students understand polyominoes, have them work on this project: There are 12 different arrangements for five squares (pentominoes); let your students find as many as they can. They might work individually or in groups. Remind students that shapes that are similar to each other (turned around and/or flipped over) are not considered different.

When your students are finished, have them pool their findings. Figure 9-5 shows the 12 possible arrangements.

Here are two further activities using polyominoes:

1) Of the 12 pentominoes, eight of them can be cut out and folded into an open box. Students can guess which ones will work, then check by cutting and folding. (Boxes I, J, K, N, P, Q, R, S.)

2) For this activity it will be necessary to cut out a model of each polyomino (as shown in Figures 9-4 and 9-5). Cut one A, one B, then shapes C through S.

These 19 polyominoes can be fitted together in different combinations to form rectangles and squares. Let your students try to form some of the following:

3 by 3 squares (E and M) or (L and G) or (C and O);
3 by 4 rectangles (B,F,I) or (A,G,K) or (B,G,M);

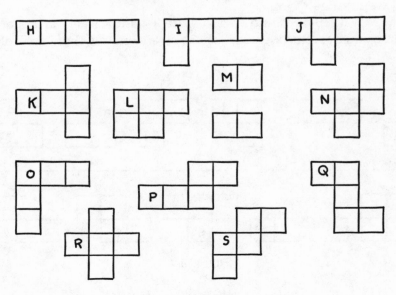

Figure 9-5

4 by 4 squares (A,B,M,S) or (A,B,I,P);
4 by 5 rectangles (J,L,M,N) or (L,M,O,Q); and
5 by 5 square (I,J,M,O,Q).

Students might be asked to draw these rectangles, then find the perimeters and areas. As the areas become larger, the rectangles become more difficult to form. Perhaps the most difficult is a 6 by 10 rectangle formed from all 12 pentominoes (H to S).

ANGLES OF A TRIANGLE

(geometry)

Your students will need to cut out several different types of triangles—equilateral, right, obtuse, and so on. The angles should be labeled. To find the sum of the three angles have your students cut out the three angles of a triangle and place them side by side with vertices touching. (See Figure 9-6.) This should be repeated with different triangles.

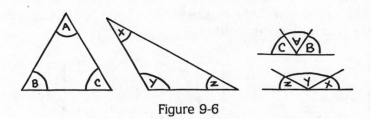

Figure 9-6

In each case your students should reach the same conclusion: The sum of the three angles of any triangle equals a straight angle—180°.

Now allow students to investigate the sum of the four angles of a quadrilateral. They should have a variety of four-sided figures of different sizes, from rectangles to parallelograms to quadrilaterals with four unequal sides. What conclusions will they reach? (The sum of the four angles of any quadrilateral equals a complete rotation—360°.)

MÖBIUS STRIPS

(geometry)

A strip of paper is joined to form a band in three different ways. When each is cut in half lengthwise surprising results occur. Möbius strips can be presented as a magic trick done by the teacher, or as an investigation into a branch of geometry known as topology, and done by the class.

Cut three long strips of paper (newspaper may be used) and join ends as shown in Figure 9-7.

(a) The first strip is joined with tape to form an ordinary band;

(b) the second strip should be given a half twist and then joined. This is the Möbius strip.

(c) The third band is given two half twists.

Make sure the ends are joined completely by placing tape as shown in the diagram.

Two questions may now be investigated: What happens when we draw a line in the middle of each strip lengthwise, going around until the line returns to its starting place? Then what happens when we cut each strip along this line?

The regular band (a) provides us with the expected. However, surprising results will occur with the Möbius strip (b) and the third band (c). When a line is drawn lengthwise on a Möbius strip (b) it must go around twice before ending, and the line will seem to be on both sides of the strip. Actually, this proves that the Möbius strip has only one side. It would therefore be impossible to color a Möbius strip red on one side and blue on the other.

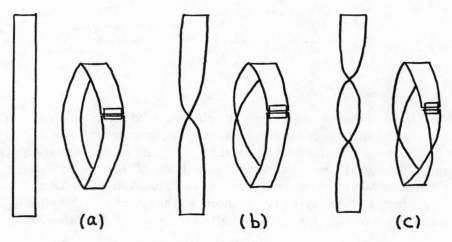

(a)   (b)   (c)

Figure 9-7

However, the most surprising results occur when we insert scissors into the strips and cut lengthwise along the drawn line. In the ordinary band (a), two separate bands are formed. But when the Möbius strip (b) is cut "in half" lengthwise, one longer band is formed rather than two separate bands. The band with the two half twists (c) does form two bands when cut, but these are interlocked.

THE FOUR-SIDED PAPER

(geometry)

An ordinary sheet of paper can be folded in a special way and made to appear as if it had four sides—each labeled differently. Years ago, similar devices were used for puzzles and children's magic tricks.

To make our four-sided paper, first fold a sheet of paper (any size) into eight rectangles and label the front as shown in Figure 9-8. Cut on the dotted lines, then turn over and label the back.

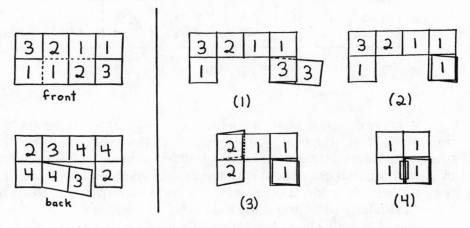

Figure 9-8

Turn front side forward and make the following special folds using the diagram as a guide: (1) Fold bottom center rectangles 1 and 2 under rectangle 3 so that a new rectangle 3 appears. (2) Fold the new rectangle 3 forward and over until side 1 appears on the lower right. (3) Next fold rectangles 3 and 1, which are on the left, back under until a second 2 appears. (4) Fold these 2s under again until all the 1 rectangles appear. Then apply transparent tape.

Now, all the 1s appear on the front, the 2s on the back. To get the third side to appear, fold the 2 side in half backwards along the center vertical fold, then open this center fold up. Do the same for the 3 side, and the last hidden side should appear.

The four sides might each be decorated with a different color or shape to make this four-sided paper look even more interesting.

WHICH IS DIFFERENT?

(logic, creativity)

In this creative approach to a familiar type of logic question students make up the questions themselves.

Start with some simple examples that you have prepared in advance. (See Figure 9-9.) A sheet of paper is divided into four rectangles. Four pictures are drawn; three of them have something in common. Students try to find the one that is different.

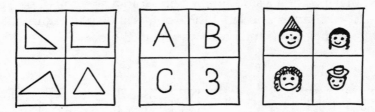

Figure 9-9

Most students will give the expected answers: the rectangle, the number 3, and the face that is frowning. However, allow for the unexpected, for example, "the letter A is different because it is made up of only straight lines." Accept all alternatives which are backed with a reasonable explanation. If colors are to be used, three pictures could be colored similarly while the third is colored differently.

Now it is your class's turn. Supply paper and have each student prepare one question of this type. Allow students, one at a time, to present their own problem to the class. How creative were your students in their questions? How logical were students in their answers?

ONE OR TWO

(logic)

This simple logic game for two players uses 10 objects that are placed in a row. Each player may in turn take either one object or two adjacent objects. Players alternate. The player who takes the last object wins. Remember, a player can take no more than two objects, and may take two only if they are adjacent (without intervening spaces). Try this sample game between player A and player B.

```
START   O  O  O  O  O  O  O  O  O  O
  (A)   O  O  O  O  —  O  O  O  O  O
  (B)   O  O  O  O     O  O  —  —  O
  (A)   —  —  O  O     O  O        O
  (B)         O  —     O  O        O
  (A)         O        —  —        O
  (B)         O                    —
  (A)   A takes the last one and wins
```

NIM

(logic)

Nim is an ancient logic game in which 15 objects are arranged as follows:

Two players alternate, and each in turn may pick up (or cross out) as many objects as he or she wishes in a single row. The object of Nim is to force the other player to take the last object.

Following is a sample game between player A and player B.

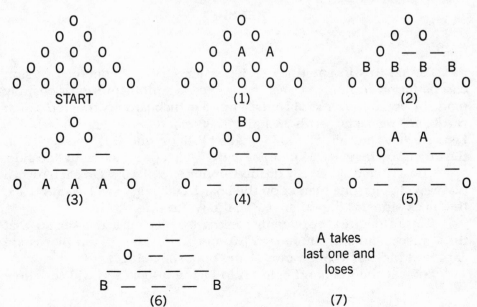

SECRET NUMBER

(logic)

A student comes up to the front of the room and writes a secret number on a piece of paper. The object of the game is to guess that number. The student must tell the class if the number has three digits, four digits, or five digits. Each digit in the number should be different; 1584 would be acceptable, but 1585 would not be.

As each guess is made, the class should be told: 1) how many digits in the guess are correct, and 2) how many of those digits are in the correct place. A list should be kept on the chalkboard, as shown in the following example. By using previous information as clues, the class should eventually be able to guess the secret number.

Here is an example of what a blackboard list might look like at the beginning of a game. Remember that the secret number is 1584.

Guesses	Digits	Places
2468	2	0
3579	1	1
3568	2	1
2451	3	0
3524	2	2
....		

REVERSI

(logic)

Reversi was invented about 100 years ago. A commercial version can be found in many stores; but it can be simply and inexpensively made by your students (36 cents). First, a gameboard is prepared, using cardboard or oaktag, with 36 squares; 36 markers will be needed. (See Figure 9-10.) They all should be made black on one side and white on the other. You may use 36 pennies with black tape attached to one side.

The markers are evenly divided among two players. On each turn, a player will place one marker on the board. One player will always place the black side up; the other player, the white side.

Players capture opponents' pieces by placing a marker so that these pieces are surrounded on two opposite sides. When pieces are captured they are flipped over to the opposite color.

Look at the left-hand column in Figure 9-10(a). You will see three

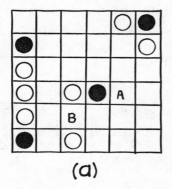
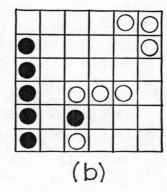

(a)   (b)

Figure 9-10

white markers surrounded by two black markers. The captured white pieces are then flipped over to their black sides.

If white, on its next move, places its piece on the square marked A, it will capture a black piece. However, black may place its next piece on square B without being captured, since the white markers had already been on the board. Captures are made only when a player actively places a piece that results in the opponent's pieces being surrounded. Thus, players are free to move between opponent's markers that have previously been placed on the board.

Figure 9-10(b) shows the results of all moves just described. Also shown is a method for capturing corner pieces. Before playing, have your class decide if they want to allow for captures along diagonal lines in addition to horizontal and vertical captures.

Note that markers may be flipped over many times in one game. *Reversi* ends when all 36 markers have been placed on the board. The player whose color appears on the most squares is the winner.

AFRICAN STONE GAME

(logic)

Thousands of years ago this game was played in Africa. It is now played in many countries under many names (for example, Mancala and Wari) and there are many variations. It is sold commercially in this country. A simple version for two players can be made with an egg carton and 36 counters (beans or pebbles).

Three counters are placed in each cup of the egg carton. The first player (player A) picks up all the counters from any one of his or her starting cups. These counters are placed one at a time into the next three cups lying counterclockwise from the starting point. [See Figure 9-11(a).] The next player takes the pieces from any of B's starting cups and places them one at a time in the three cups lying counterclockwise from the starting point. [See Figure 9-11(b).]

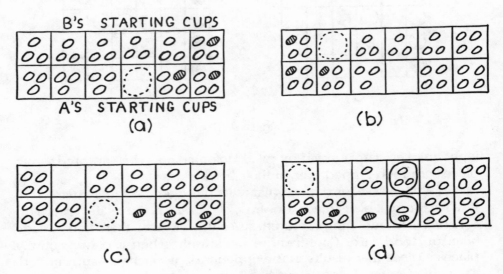

Figure 9-11

The game continues with players alternating. Players try to move pieces so that last counter lands in a cup that contains only one counter. The player wins these two counters, and, as a bonus, also takes all the pieces in the opposite cup. For example, after player A has moved as shown in diagram (c), player B takes the four counters from the upper left-hand cup and distributes them. [See diagram (d).] Since the last piece landed in a cup that had contained only one counter, the player wins these two counters plus the counters in the opposite cup. These five counters are circled in diagram (d), and will be taken from the board by player B.

The game ends when all six of one player's starting cups are empty, and it is that player's turn to move. The players will then count their winning pieces, and the player having the most counters is the winner.

MATH CREATIVITY TEST

(creativity)

To make up a math creativity test, you might include several open-ended problems similar to those following.

I. TWO SAMPLE QUESTIONS TO BE DONE WITH CLASS:

1. Write many different number sentences (equations) using numbers 1,2,3,4,5,6. (Answers could include $1+2=3$, $3-2=1$, $2\times3=5+1$, $4\div2=3-1$, and $1+2+3=6$)

2. a) How is an inch different from a centimeter?

b) How is an inch similar to a centimeter? (Answers could be, a) an inch is larger, b) both are units of measure.)

II. SIX TEST QUESTIONS TO BE DONE BY STUDENTS:

1. Using only these six numbers: 1,2,5,8,10,12 write as many different number sentences as you can. (Each number should be used only one time in each equation.)

2. How many different ways can you divide a square into four equal parts? Illustrate each answer. *Note to teacher:* If necessary, illustrate for your class how to divide a circle into four equal parts using two diameters.

3. How are the numbers 10 and 100 different? How are they alike? Give as many answers as you can to both questions.

4. How is a triangle different from a square? How is a triangle similar to a square? Give many different answers. *Note to teacher:* Use equilateral triangle in this problem if your students are familiar with its meaning.

5. What is your favorite number? Give many reasons for your answer.

6. What is your favorite geometric shape? Give many reasons for your answer.

Students should be given ample time for a creativity test. If six questions such as the previous ones were used, 30 minutes should be allowed. In creative thinking it is sometimes best to leave a problem incomplete and come back to it later with fresh insight.

Since all these questions are open-ended, there is no correct score for this test. Rather, the object is to see how many different answers a student can have. Also, we are interested in any completely unique solutions that may appear.

YOUR OWN NUMBER SYSTEM

(creativity)

This activity is not for all students, but for higher ability students, writing their own number system should be an interesting project.

First, students need to be familiar with a few different number systems such as our own base 10 system, Roman numerals, and the Egyptian hieroglyphic number system (mentioned in Chapter 7). Although it is not necessary, another base system might be shown to the students.

Now students may be asked if they would like to invent their own number system. This could be done individually or in groups.

The new number system could parallel our own number system with only the symbols (digits) changed, or it could use a different base, or it could be a repetitive system such as that which the Egyptians used.

Can students write numbers such as 65 and 479 in another number system? Have them add 12 + 38. Can they multiply with the number system? What is the largest number they can write using their new symbols?

Figure 9-12 is an example of a number system created by a student. At first this student did not have a symbol for zero, but when adding 12 + 38 she realized the need and added the zero. Note that since this number system uses base 10, the symbol for 10 is not really needed.

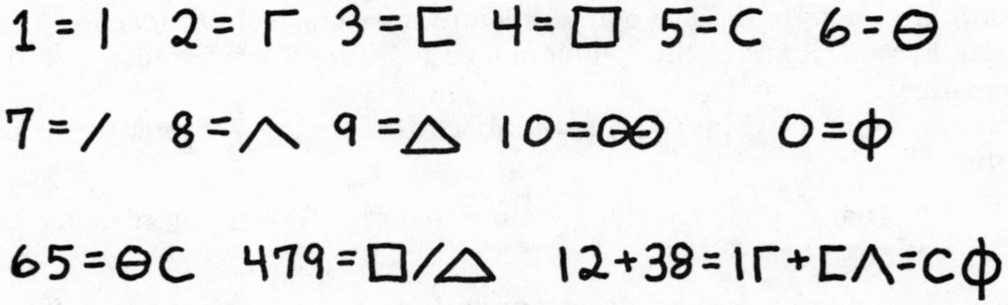

$$1 = |\quad 2 = \Gamma \quad 3 = \mathsf{C} \quad 4 = \square \quad 5 = \mathsf{C} \quad 6 = \ominus$$

$$7 = / \quad 8 = \wedge \quad 9 = \triangle \quad 10 = \infty \quad\quad 0 = \phi$$

$$65 = \ominus\mathsf{C} \quad 479 = \square/\triangle \quad 12 + 38 = |\Gamma + \mathsf{C}\wedge = \mathsf{C}\phi$$

Figure 9-12

GEOMETRY DESIGN CONTEST

(creativity, geometry)

Although this is a contest, everyone can win. First ask students to draw designs using only the basic geometric figures. If necessary, review these figures with your class. These shapes may be arranged and colored in any way.

After all students hand in their designs, pick a few winners in each category: most colorful, most interesting, neatest, most creative, nicest arrangement, and so on. The idea here is for each student to be a winner in at least one category.

Finally, all pictures should be displayed in your classroom.

STUDENT-MADE MATH GAMES

(creativity)

After your students have played many of the math games in the earlier chapters, they might be interested in inventing their own math games.

Divide your class into groups; each group should work together to make one math learning game which they will present to the class. However, if your students are quite independent and creative they may wish to work alone to create their own individual math games. In either case, tell students what the math objectives of the games might be (for example, practice addition combinations, review multiplication by 9, and so on).

Some students will need more help or advice than others. Tell your students that they may invent a completely new game or use a variation of a familiar game. Here are some variations you can suggest if necessary:

1) tic-tac-toe with math questions;

2) a form of Bingo using math problems (why not call it MATHO?);

3) make cards similar to flash cards and decide on rules for scoring;

4) a math version of a spelling bee;

5) a board game with markers that move around the board involving math questions.

But best of all, encourage students to use their imaginations and create their own games. All games should then be shared with the class, played, and enjoyed.

CHAPTER 10

Deal the Cards:
an Assortment of Card Games
Designed to Teach
and Review Mathematics

Card games have been around for thousands of years; no one knows just how many different types of card games have been invented over the centuries. Here is a collection of new and unique card games. All of these have been used with children in a way to involve them with mathematics, from the simple task of ordering numbers to a complex game containing multiplication of a three-digit number.

"Ten Pick-Up" and "Leave Four" are favorites of students—both those who are presently learning their addition combinations and those who have long ago mastered these facts. "Multiplication Cross-Out" is a game for one or more students in which individual choices can greatly affect the outcome.

In "Card Equations" all operations may be combined; "Seven and One-Half" uses addition with fractions; and "Decimal War" involves understanding and comparing decimals. An unusual card game is "The Sixth Card"—a game in which patterns, logic, and reasoning are used.

In addition to these card games developed by the author, there are many traditional games, such as casino, which involve mathematics to various degrees. By choosing the appropriate card games for your students, learning can be combined with play.

ONE TO TEN

An easy game or activity for students involving recognizing and ordering numbers. Students arrange cards in numerical order; may be done individually or played as a competitive game.

Preparation: Remove picture cards from deck and sort remaining 40 cards into suits. (Optional: Using tape, replace each "A" on the aces with a "1.")

Directions:

Each student is given 10 cards of one suit, which the student arranges in order from one (or ace) to 10, in a left to right sequence.

To make a competitive game, all students start at the same time. The first student to finish is the first winner, the second is the second winner, and so on. All students correctly arranging cards should be called winners.

Note: Cards can be very useful for students as they learn to count and recognize numerals. Each card contains both the written numerals and the corresponding number of symbols (hearts, clubs, and so on). Young students may wish to make up their own games with a set of cards.

TEN PICK-UP

An addition card game for two to four players. Students turn over cards as they look for pairs that have a sum of 10. A variation is suggested for sums of 13.

Preparation: All picture cards should be removed from the deck. The remaining 40 cards are spread out face down across the playing area so that no two cards are touching.

Directions:

The first player turns over two cards one at a time. If the sum of these two cards is 10 (for example, 4 and 6, or ace (1) and 9), the player picks up and keeps both cards.

If the sum is not 10, the two cards should be turned over and placed face down in their original positions, but only after the other players have seen the face value of the two cards (remembering the positions of these cards will be useful). However if one of the two cards should be a 10, the player will keep the 10 but return the other card.

Whenever a player picks a pair whose sum is 10, he or she continues to play. When two cards are chosen which do not have the correct sum, the next player (usually the student on the left) has a turn.

The play continues until all the cards have been picked up. The student who then has the most cards is the winner.

Note: Cards have the advantage that if students are unsure of the sum, the students can count the symbols—for example, seven diamonds plus three clubs make a total of 10.

Variation: Thirteen Pick-Up: Give the kings a value of 13, the queens a value of 12, and the jacks a value of 11. Use all 52 cards. This time pairs that equal 13 (for example 6 and 7, 2 and jack) or a king would be kept.

LEAVE FOUR

This is an interesting solitaire game where simple addition (or subtraction) combinations are used to remove cards. Object is to keep finding combinations and removing cards until only four cards are left. While only one student can play at a time, other students might enjoy watching (and giving advice).

Preparation: 40 cards, ace through 10, will be used. All picture cards should be removed; the ace will be used as a one.

Directions:

All 40 cards are dealt face-down into eight piles. Five cards should be in each pile. The eight piles are best arranged in two rows of four.

To begin the game, a player turns over the top card on each pile. The top eight cards might look like this:

5	6	1	10
9	2	3	2

For the rest of the game students will be removing addition (or subtraction) combinations, three cards at a time. In the above example, the student could remove the 9, the 1, and the 10. $(9 + 1 = 10)$

5	6	X	X
X	2	3	2

Now there are three piles with only face-down cards. Turn over the top card in each of these three piles. It might now look like this:

5	6	7	4
8	2	3	2

The student continues to remove combinations, 2 + 3 = 5, or 6 + 2 = 8, 3 + 4 = 7, and so on. Every time three cards are removed from their piles, the top cards on these three piles should be turned over.

Later in the game all the cards in some piles will be removed, leaving an empty space:

4	10	3	2
5		7	8

When this happens, the student may move any one of the remaining face-up cards into this space. In the above example, suppose we move the 10 to fill the empty space:

4	X	3	2
5	10	7	8

Now we can turn over the face-down card that had been under the 10:

4	9	3	2
5	10	7	8

The game continues with the student removing combinations (three cards at a time), turning over cards, and filling in empty spaces. Play continues until no more three-card combinations can be made.

A player is considered a winner if he or she is able to remove almost all of the cards and leave only four. (Hence the name "Leave Four.")

Let's see how a game might end. At this point all face-down cards have been turned over and there is one empty space:

5	7		8
2	9	8	2

The player now removes a 2, a 7, and a 9 (2 + 7 = 9), leaving only four cards, and is a winner.

Note: As with all card games, some luck is involved; but by using the following strategies, a player can win more than half the time. Sometimes a student will be a super-winner, leaving only one card.

1. When there is a choice of combinations, always choose the one

that contains the largest number. (Note that in the sample game shown, the player's first choice was $9 + 1 = 10$ rather than $2 + 1 = 3$.)

2. Always turn over face-down cards when they become exposed and fill in all empty spaces promptly so that eight face-up cards will always be showing.

COLOR MATCH

A card game involving addition of one- and two-digit numbers. When cards of the same color have been picked, the student receives a score equal to the sum of the cards. These scores are then added together for the final score. Throughout the game, students have the opportunity to "take a chance," where they may win extra points or lose them.

Preparation: Remove picture cards from the deck. Deal out 16 cards face down in a 4 by 4 array. Students will need paper and pencil for scoring.

Directions:

The first player will pick up two cards. Are they the same color? If not, cards should be returned face down in their original position.

However if cards are a color match (both red or both black) then the student can keep the two cards and receive a score equal to their sum. Or the student can choose to "take a chance," and pick up a third card.

In this case, the third card must be the same color as the first two to receive a score. A successful player would get a score equal to the sum of the three cards, while an unsuccessful player (whose third card did not match) would return all three cards and receive no score.

Players take turns and whenever a score is received, the student adds it to the previous score. After all possible color pairs have been picked up, the student with the highest total score wins.

Example: A player has picked up two cards. They turn out to be the 5 of diamonds and 7 of hearts. There are now three possibilities:

1. Player keeps these two cards and receives a score of 12. This player's turn would now be over.

2. Player decides to take a chance and turns over the 3 of clubs. All three cards must be returned, and the player receives no score on that turn. (Other players try to remember where these cards have been placed.)

3. Player takes a chance and turns over the 8 of diamonds. This student will receive 20 points which will be added to the previous score. The next player now has a turn.

TWENTY-ONE TAKES ALL

A group of students may play this card game, in which the face value of the cards are added orally until a sum of 21 is reached. Can be played with three to five players, but is best when five play.

Preparation: The picture cards are removed from the deck and each student receives seven cards.

Directions:

First player puts a card down and announces its face value (for example "4"). The next player, on the left, places another card down; this time announcing the sum of the two cards. (If the second card is a 5, the player calls out "9.") The next student puts down a card and announces the sum of all three cards (an 8 would make the sum 17).

The player who places a card making the total exactly equal to 21 takes the whole pile. But the total may not go over 21. In the example, the fourth student may put down an ace (1), 2, 3, or 4; but not any higher card.

If a player does not have a card small enough to make the total equal to or less than 21, the player then passes. When 21 is reached, the pile is picked up by that player. Winning cards are put aside, to be counted at the end of the game. The next player then puts down a card and begins a new pile.

The game continues in this manner with players picking up "21" piles. When no player has a card small enough to place on a pile (keeping the sum 21 or under), the game is over and the student who has picked up the most cards is the winner.

MULTIPLICATION CROSS-OUT

In this game each student has a list of digits from 0 to 9. The face value of two cards are multiplied, and then the digits in their product are crossed out. Object is to cross out as many digits (0 to 9) as possible in five turns. Choices are involved, and students' skills (and a little luck) will determine who wins. It can be played by one student or a group.

Preparation: Each student will need a pencil and paper with the digits 0 to 9 listed. Picture cards are to be removed and 10 cards given to each player. (A multiplication table should be available for students.)

Directions:

Players lay out their 10 cards so that all 10 can be seen. Player selects two cards, then multiplies their face value to find the product. These two cards are now removed, and the digits of this product are crossed out. A list should be kept of all factors and products used, in case it is later necessary to check work.

For one player: After laying out the 10 cards, the player picks up five pairs of cards, one pair at a time. Each time the student will cross out the digits of the product. Can the player cross out all 10 digits? If not, deal 10 more cards and try again. Depending on the player's cards, it may not be possible. A good score would consist of crossing out at least seven digits. See examples below.

For a few students: Players work independently with the 10 cards they have been dealt. After each student has selected five pairs, the student with the most digits crossed out wins the game. See examples.

Examples:

Player A is dealt: ace (1), 2, 2, 3, 4, 5, 6, 6, 7, 9.   0
first pair: 1 and 6. Since 1 × 6 = 6, cross out 6.   1̶
second pair: 2 and 7. Since 2 × 7 = 14, cross out 1 and 4.   2
third pair: 6 and 9. Since 6 × 9 = 54, cross out 5. The 4 was already   3
                    crossed out.   4̶
fourth pair: 2 and 4. Since 2 × 4 = 8, cross out 8.   5̶
fifth pair: 3 and 5. Although 3 × 5 = 15, both the 1 and the 5 were 6̶
                    already crossed out.   7
                                                                         8̶
                                                                         9

After choosing these five pairs, Player A has crossed out only five digits. This is not a good score. By choosing different pairs, the same player could have crossed out seven digits:

		0̶
		1
first pair:	1 × 9 = 9	2̶
second pair:	5 × 6 = 30	3̶
third pair:	2 × 4 = 8	4̶
fourth pair:	7 × 6 = 42	5̶
fifth pair:	3 × 2 = 6	6
		7
		8̶
		9̶

Challenge: If you were dealt: 2, 3, 5, 6, 7, 8, 8, 9, 9, 10; how would you pair these cards so that all digits (0 to 9) could be crossed out?

MULTIPLYING CARDS

A difficult game involving multiplication of a three-digit number by a one-digit number, where cards are used as the digits—no pencil or paper. It is for two teams or two players. Also, a dealer will be needed. For team play, students must cooperate with each other or their team will not be very successful. (A solitaire challenge is also provided.)

Preparation: A dealer will remove all picture cards, and separate the remaining 40 cards into two piles; one containing the black cards and the other containing the red cards.

The dealer, using the pile of black cards, will deal out four face-up cards and 10 face-down cards to the first player (or team). The second player will be dealt the red cards, four face up and 10 face down.

If any of the face-up cards is a "10," the dealer should replace it with another card of the same color.

Directions:

Players may now look at their face-down cards. We will call these 10 cards the "product cards." These cards may contain a "10." Since the numeral 10 contains two digits, 1 and 0, in this game the "10" card may be used as either a "1" or a "0." The ace will be used as a 1; all other cards will count as their face value.

The players should now arrange their four face-up cards into the form of a multiplication problem with three digits multiplied by one digit.

If the four face-up cards are 4, 3, 5, and 9; player might make any of the following arrangements:

$$
\begin{array}{ccccc}
459 & 435 & 394 & 945 & 493 \\
\times 3 & \times 9 & \times 5 & \times 3 & \times 5 \\
\end{array}
$$

Now to win the game, the player (or team) must place the proper "product cards" down to form the correct product for the problem which they have laid out. If the 10 product cards do not contain the correct digits for a particular product, players may change around the order of the four face-up cards—thereby forming a new problem (see example below).

The first player or team to successfully complete a product wins. (The dealer should check the multiplication.)

If after a certain time limit, no one has found an arrangement that works for the cards available, the dealer will stop the game, collect the cards and redeal.

Example: A team has been dealt the following red cards:

    face up: 3, 4, 5, 8
    face down: A, 3, 4, 5, 6, 6, 7, 9, 10, 10
    The team first tried: 358
                             $\times 4$

But this was a problem: $4 \times 8 = 32$ and the team did not hold a "2" card among their 10 product cards.

So the team then tried: 385
                           $\times 4$

This proved to be successful. The team worked on the product from right to left. Since $4 \times 5 = 20$, the team placed a "10" card to represent the 0, and carried the 2 to the 10s column (regrouping). Now $4 \times 8 = 32$, and $32 + 2 = 34$. Next, a 4 was taken from the product cards and placed down. The 3 to be carried (regrouped) was remembered. Continuing, $4 \times 3 = 12$ and $12 + 3 = 15$. A 5 card was placed in the product and finally an ace was used to represent 1.

The final results: 385
                      $\times 4$
                    ⎯⎯⎯⎯
                      1540

Challenge: Here is an interesting solitaire game for one player. Student is dealt 10 cards. The object is to form a multiplication problem using any four cards (three-digits times one-digit), so that the answer can be found among the remaining six cards.

Example: with A, 2, 3, 3, 4, 6, 7, 7, 9, 10, a player can form

        317   or   773
       $\times 2$ $\times 4$
      ⎯⎯⎯⎯   ⎯⎯⎯⎯
        634   3092

CARD EQUATIONS

A card game for two to eight players. Using addition, subtraction, multiplication, and division, students form their own equations. The more cards that are used in the equation, the higher the score.

Preparation: A full deck of cards can be used or the picture cards may be removed. If picture cards are used, Jack = 11, Queen = 12, King = 13. Paper and pencil will be needed to keep score.

Directions:

Each player is dealt five cards. Student tries to make equations, using the cards that have been dealt.

If a player's cards are 2, 3, 4, 6, 10, this student could make $4 + 6 = 10$, or $2 \times 3 = 6$, or $3 \times 4 = 10 + 2$. The score is determined by the number of cards used:

Score if 3 cards are used = 3 points,
4 cards are used = 4 points,
5 cards are used = 5 points.

Each player in turn lays down the cards for the equation he or she wishes to make, and calls out the equation. In the above example the student would lay out the 3, the 4, the 10 and the 2 and state "3 times 4 equals 10 plus 2." If the other players agree that the equation is correct, the student receives four points.

Actually, there was an equation possible using five cards. Can you find it? (Answer: $(6 \times 3) - (4 \times 2) = 10$.)

If the student cannot make an equation, the student would receive no points for that turn.

When the round is over, all the cards would be collected and new cards dealt. After five rounds, players add the scores. The highest score wins.

Examples:

Player A's cards: 1,2,4,5,8   Equation: $2 \times 4 = 8$ (3 points)
Player B's cards: 1,3,6,7,10   Equation: $10 + 7 + 1 = 6 \times 3$ (5 points)
Player C's cards: 3,4,8,9,Q   Equation: $9 \div 3 = 12 \div 4$ (4 points)
Player D's cards: 2,5,8,Q,K   Equation: $5 + 8 = 13$ (3 points)

SEVEN AND ONE-HALF

A card game similar to Blackjack, but using addition of fractions.

Preparation: From a deck of cards, remove the 6s, 7s, 8s, 9s, and 10s; 32 cards remain. They will be given the following values:

A to 5 = face value
King = $\frac{1}{2}$
Queen = $\frac{1}{3}$
Jack = $\frac{1}{4}$

Keep chart of these values available for reference. Also have paper and pencil handy if needed.

Directions:

Dealer gives each player one card face down. Each player looks at the card and then decides if he or she wants another card placed face up (as in Blackjack).

Players may choose to stop at one (face down) or receive one at a time as many face-up cards as they wish.

The student who has a sum closest to $7\frac{1}{2}$ is the winner, and becomes the dealer for the next game.

It is all right to go over $7\frac{1}{2}$. For example, if one player has A, 5, K = $6\frac{1}{2}$, and another has 3, 4, Q, Q = $7\frac{2}{3}$, then the player with $7\frac{2}{3}$ wins.

Note: Which is closer to $7\frac{1}{2}$: $6\frac{3}{4}$ or $8\frac{1}{3}$? Check by using subtraction. $7\frac{1}{2} - 6\frac{3}{4} = \frac{3}{4} = \frac{9}{12}$ while $8\frac{1}{3} - 7\frac{1}{2} = \frac{5}{6} = \frac{10}{12}$. Therefore $6\frac{3}{4}$ is closer.

DECIMAL WAR

Played just like the traditional card game War, but this time decimals are used. Students must decide which is larger, 6.7 or 7.6, 5.10 or 5.8.

Preparation: Remove the picture cards and divide the remaining 40 cards into two piles, one containing the red cards, the other, the black cards. Both players should be dealt 10 black cards and 10 red cards. The black cards represent whole numbers; the red cards will be tenths.

Directions:

Each player places his or her cards into two face-down piles: one for their black cards; the other for their red cards.

Players may sit opposite or next to each other. Both students simultaneously turn over the top card of their black pile with their left hand and the top card of the red pile with their right hand.

If one student had a black 5 and a red 2, this would be read as 5.2 (5 and 2 tenths). If the other player has a black 3 and a red 8, 3.8 would be formed.

The player with the highest number wins all four cards. In the example, the student with 5.2 wins these four cards. The game continues in this manner, with players winning four cards at a time.

The winning black cards are placed face down at the bottom of the player's black pile, while the winning red cards are placed face down on the bottom of the red pile.

The game is over when one player finally wins all 40 cards.

Note: Before your students play, let them practice comparing decimals. Which is larger: 8.9 or 9.8? 3.7 or 4.1? 6.5 or 6.10? Discuss with students that 6.10 equals 6.1, therefore 6.5 is larger than 6.10.

THE SIXTH CARD: A LOGIC GAME

A student will arrange six cards in some order. The other students will be shown only the first five cards in the series. They will try to discover the rule used to order these cards, and to guess the sixth card.

Preparation: One student will be chosen to be the "rule maker." He or she will be given a deck of cards and allowed some time to think up a rule for ordering the cards.

Directions:

The rule maker selects and arranges six cards that illustrate his or her rule. The first five cards are placed face up (in a left to right sequence), and the sixth card is placed face down.

The rule may be based on the color of the cards, the number on the cards, or even the suits. Here are some ideas: alternating colors, numerical order (forward or backward), repeating patterns, and so on. In this game an ace can count as a 1, or it can come after a King (as in sequence J,Q,K,A).

After cards have been laid out, other students try to guess the rule and the sixth card, then the sixth card is turned over. The student who was first to guess the rule is given the deck of cards and becomes the new rule maker.

In the following illustrations, these abbreviations will be used: H = hearts, D = diamonds, S = spades, C = clubs, X = face-down sixth card. See if your students can guess these rules, and predict the number and/or color of the sixth card:

1) 8D   7H   6S   5D   4C   **X**
Rule: backward sequence   X = a 3 (any suit)

2) 2H   8S   KH   3S   5H   **X**
Rule: alternating colors   X = a black card (spades)

3) JC   QD   KH   JD   QS   **X**
Rule: repeating pattern,
Jack, Queen, King   X = a King

4) 3D   5H   7H   9D   JD   **X**
Rule: every other number
sequence, all red   X = a red King.

Subject Index*

*PLEASE NOTE: This index does not include the games and activities in Chapter 6, each of which can be used for a variety of purposes—addition, subtraction, multiplication, division, etc.

DATE DUE

1/29/85	NOV 0 5 2010	
10/2/85		
MAR 1 7 1987		
10/12/87		
NOV 0 2 1989		
MAR 0 9 1991		
SEP 2 9 1991		
MAR 1 1 1992		
5-10-92		
MAR 0 9 1993		
MAR 2 3 1993		
12-20-93		
MAR 1 0 1994		
NOV 1 9 1994		
NOV 2 5 1995		
FEB 0 8 1996		
OCT 2 2 2000		

HIGHSMITH 45-102   PRINTED IN U.S.A.